CW01502049

VOLUME TWO

ADVANCES IN
# CHEMICAL POLLUTION, ENVIRONMENTAL MANAGEMENT AND PROTECTION

Sustainable Use of Chemicals in Agriculture

ADVANCES IN
# CHEMICAL POLLUTION, ENVIRONMENTAL MANAGEMENT AND PROTECTION

## Sustainable Use of Chemicals in Agriculture

Edited by

**ETTORE CAPRI**
*Università Cattolica del Sacro Cuore,*
*Istituto di Chimica ed Ambientale,*
*Piacenza, Italy*

**ANNE ALIX**
*Corteva Agrisciences™,*
*the Agriculture Division of DowDupont™,*
*Abingdon, United Kingdom*

ACADEMIC PRESS
An imprint of Elsevier

Academic Press is an imprint of Elsevier
50 Hampshire Street, 5th Floor, Cambridge, MA 02139, United States
525 B Street, Suite 1800, San Diego, CA 92101–4495, United States
The Boulevard, Langford Lane, Kidlington, Oxford OX5 1GB, United Kingdom
125 London Wall, London, EC2Y 5AS, United Kingdom

First edition 2018

Copyright © 2018 Elsevier Inc. All rights reserved.

No part of this publication may be reproduced or transmitted in any form or by any means, electronic or mechanical, including photocopying, recording, or any information storage and retrieval system, without permission in writing from the publisher. Details on how to seek permission, further information about the Publisher's permissions policies and our arrangements with organizations such as the Copyright Clearance Center and the Copyright Licensing Agency, can be found at our website: www.elsevier.com/permissions.

This book and the individual contributions contained in it are protected under copyright by the Publisher (other than as may be noted herein).

**Notices**
Knowledge and best practice in this field are constantly changing. As new research and experience broaden our understanding, changes in research methods, professional practices, or medical treatment may become necessary.

Practitioners and researchers must always rely on their own experience and knowledge in evaluating and using any information, methods, compounds, or experiments described herein. In using such information or methods they should be mindful of their own safety and the safety of others, including parties for whom they have a professional responsibility.

To the fullest extent of the law, neither the Publisher nor the authors, contributors, or editors, assume any liability for any injury and/or damage to persons or property as a matter of products liability, negligence or otherwise, or from any use or operation of any methods, products, instructions, or ideas contained in the material herein.

ISBN: 978-0-12-812866-4
ISSN: 2468-9289

For information on all Academic Press publications
visit our website at https://www.elsevier.com/books-and-journals

Working together
to grow libraries in
developing countries

www.elsevier.com • www.bookaid.org

*Publisher:* Zoe Kruze
*Acquisition Editor:* Jason Mitchell
*Editorial Project Manager:* Shellie Bryant
*Production Project Manager:* Vignesh Tamil
*Cover Designer:* Victoria Pearson

Typeset by SPi Global, India

# CONTENTS

Contributors                                                                                    vii
Series Editor's Preface                                                                          ix
Preface                                                                                          xi

1. **Modern Agriculture in Europe and the Role of Pesticides**                                    **1**
   Anne Alix and Ettore Capri

   1. Introduction                                                                               2
   2. (EC) Regulation No. 1107/2009 on the Placing of Pesticides
      on the Market                                                                              4
   3. Directive 2009/128 (EC): A Framework to Further Secure Pesticide
      Use and Application Toward Sustainability                                                  11
   4. Effectiveness of the Regulatory Framework in the Context
      of Sustainable Use of Pesticides                                                           14
   5. Risk Management: Toward a Sustainable Agriculture at a Larger Scale                         17
   6. Conclusions                                                                                20
   References                                                                                    21
   Further Reading                                                                               22

2. **The Multiactor Approach Enabling Engagement of Actors
   in Sustainable Use of Chemicals in Agriculture**                                               **23**
   Els Belmans, Paul Campling, Elien Dupon, Ingeborg Joris, Eva Kerselaers,

   Saskia Lammens, Lies Messely, Ellen Pauwelyn, Piet Seuntjens,

   and Erwin Wauters

   1. Introduction                                                                               24
   2. General Methodology                                                                        25
   3. The Multiactor Approach                                                                    27
   4. Water Governance                                                                           34
   5. Participatory Monitoring                                                                   49
   6. Best Management Practices                                                                  50
   7. Collaborative Software Tools                                                               50
   8. Results From Example Case Studies                                                          51
   9. Conclusions                                                                                59
   Acknowledgments                                                                               59
   References                                                                                    59

**3. Certification and Added Value for Farm Productions**                          **63**

Pieter Ravaglia, Jacopo Famiglietti, and Fiamma Valentino

1. Introduction                                                                              65
2. Agriculture in Europe                                                                     66
3. Quality in the Agro-Food Sector                                                           68
4. Origin of Sustainable Consumption                                                         69
5. 360° Quality: Classification of Public and Private Certification Initiatives              72
6. Focus on Geographical Indications                                                         91
7. Multiplication of Environmental and Sustainability Labels                                 93
8. European Consumer Behavior                                                                94
9. Conclusions and Future Developments                                                       95
Annex I                                                                                      98
Annex II                                                                                    101
References                                                                                  106

**4. The Role of Research, Communication, and Education
for a Sustainable Use of Pesticides**                                             **109**

Maura Calliera and Alba L'Astorina

1. Introduction                                                                             110
2. Sustainable Agriculture and the Role of Sustainable Development
   European Policies on Pesticide Use Education                                             112
3. State of the Art: Lessons Learnt From Research Projects Aimed
   to Foster Education in Sustainable Use of Pesticide                                      114
4. Education and Communication for a More Integrated System
   of Knowledge                                                                             119
5. The RRI Approach for a Sustainable Use of Pesticides in Agriculture                      125
6. Conclusions                                                                              129
References                                                                                  130

**5. Practical Implementation of the Principles of the Sustainable
Use of Pesticides**                                                               **133**

Manpriet Singh, Vasileios P. Vasileiadis, and Anaïs Junger

1. Introduction                                                                             134
2. Policies for Sustainable Agriculture                                                     135
3. EU Policy Design and Enabling Conditions for Implementation                              140
4. Case Studies                                                                             143
5. Conclusion                                                                               160
References                                                                                  162
Further Reading                                                                             164

# CONTRIBUTORS

**Anne Alix**
Corteva Agrisciences™, the Agriculture Division of DowDupont™, Abingdon, United Kingdom

**Els Belmans**
Flanders Research Institute for Agriculture, Fisheries and Food, Social Sciences Unit, Merelbeke, Belgium

**Maura Calliera**
Department for Sustainable Food Process, Dipartimento di Scienze e tecnologie alimentari per una filiera agro-alimentare sostenibile, Università Cattolica del Sacro Cuore, Piacenza, Italy

**Paul Campling**
Flemish Institute for Technological Research, VITO, Environmental Modeling Unit, Mol, Belgium

**Ettore Capri**
Università Cattolica del Sacro Cuore, Istituto di Chimica ed Ambientale, Piacenza, Italy

**Elien Dupon**
Inagro, Rumbeke-Beitem, Belgium

**Jacopo Famiglietti**
Independent author

**Ingeborg Joris**
Flemish Institute for Technological Research, VITO, Environmental Modeling Unit, Mol, Belgium

**Anaïs Junger**
Syngenta France SAS, Guyancourt Cedex, France

**Eva Kerselaers**
Flanders Research Institute for Agriculture, Fisheries and Food, Social Sciences Unit, Merelbeke, Belgium

**Saskia Lammens**
Flanders Environment Agency, Buitendienst IJzer, Leie en Brugse Polders, Oostende, Belgium

**Alba L'Astorina**
Institute for Remote Sensing of Environment, National Research Council, Milano, Italy

**Lies Messely**
Flanders Research Institute for Agriculture, Fisheries and Food, Social Sciences Unit, Merelbeke, Belgium

**Ellen Pauwelyn**
Inagro, Rumbeke-Beitem, Belgium

**Pieter Ravaglia**
Department for Sustainable food process, Università Cattolica del Sacro Cuore, Piacenza, Italy

**Piet Seuntjens**
Flemish Institute for Technological Research, VITO, Environmental Modeling Unit, Mol; Ghent University, Ghent; University of Antwerp, Institute of Environment and Sustainable Development, Wilrijk, Belgium

**Manpriet Singh**
Syngenta Crop Protection AG, Basel, Switzerland

**Fiamma Valentino**
DG for Sustainable Development (SVI), Italian Ministry for the Environment Land and Sea, Rome, Italy

**Vasileios P. Vasileiadis**
Syngenta Crop Protection AG, Basel, Switzerland

**Erwin Wauters**
Flanders Research Institute for Agriculture, Fisheries and Food, Social Sciences Unit, Merelbeke, Belgium

# SERIES EDITOR'S PREFACE

I am pleased to introduce the second volume of this new Elsevier series on *Advances in Chemical Pollution, Environmental Management and Protection*. The topic of this new book is: *Sustainable Use of Chemicals in Agriculture.*

I agree with the sentence written by A. Alix and E. Capri in Chapter 1 "Agriculture is in constant evolution, and modern agriculture relates to a range of revolutionary steps that aimed at improving production, reducing workload and work penibility and since the last few decades actively improve human and environmental safety. Alongside changes in practices at the producer's scale, this evolution has been echoed in the regulatory area, at the national level and further at the European level, and in the world. Particular attention was given to plant protection products (PPP), or pesticides". Later they also emphasize the meaning of sustainable use of pesticides: *to reduce risks and impacts of pesticide use on human health and the environment and at encouraging the development and introduction of integrated pest management and of alternative approaches or techniques to reduce dependency on pesticide use."*

Since the mid-1950s the use of pesticides has grown continuously every year so that the total amount of pesticide active ingredients in use are now around 2.5 million pounds per year. Pesticides, together with fertilizers, play a central role in agriculture and contribute to enhanced food production worldwide. The need for food is directly related to population growth. Annual global net population growth is 80 million and the population of the world is expected to be 9 billion people by 2030. This issue has been clearly raised by J.R. Beddington and co-workers in a Science paper published in 2010 (H. Charles J. Godfray, J.R. Beddington, et al., *Science*, 12 February 2010, vol. 327, no 5967, pp. 812–818). A "perfect storm" of food shortages together with water scarcity and insufficient energy resources is expected to occur by then. It is expected that food will play a key role due in increasing demand. So pesticides will be needed to increase food production mainly due to the need for growing more crops more efficiently.

Following the ban of DDT and other organochlorinated pesticides in the mid-1970s and its regulation by the International Convention on Persistent Organic Pollutants, a new generation of pesticides is being used. The main characteristic of these new pesticides is that they are moderately hydrophobic, and many scientists refer to them as new, modern, and/or polar pesticides. Although this new chemical group of pesticides is much less harmful than organochlorinated pesticides, there remains the need to check that they will not affect ecosystems, or trigger diversity losses affecting aquatic species, terrestrial species nor display endocrine disruption.

Since there is a general agreement that pesticides need to be reduced and need to be more sustainable I have decided to invite my old friend and former colleague at the EFSA PPP Panel Ettore Capri to edit this new book in the series. This book contains state-of-the-art contributions compiled in five chapters covering all different aspects of sustainable use of chemicals in agriculture, including multiactor approach management, certification and added value for farm production, the influence of research, communication and education as key factors to improve the use of eco-friendly pesticides, and practical implementation principles of the sustainable use of pesticides.

That being said, this book offers a unique opportunity to better understand and move forward toward a sustainable use of chemicals in agriculture. This will be a great challenge due to the need of increasing food production in the years to come. The book can be used as an academic text and as a reference book for those working in agriculture, water, and environmental agencies both governmental and private as well as by all technical people, farmers, agroindustry, and all stakeholders of the agro sector. Finally I would like to thank both coeditors and all coauthors of this book, well-known experts, for their time and efforts in preparing this excellent and useful book on the sustainable use of chemicals in agriculture.

D. BARCELO
Barcelona and Girona, Spain, April 12, 2018

# PREFACE

Where everyone can easily agree that farmers need to protect their crops from diseases, herbivorous species, concurrent weeds, or other macro- or microorganisms that can affect the growth and yields and need control, the use of pesticides as a crop protection treatment is more and more challenged by the public and even at political level. Yet, with the increasing pressure to produce food with high quality and insufficient amounts at an affordable price, farmers need protection tools they can rely on and pesticides, or plant protection products (PPP) remain reliable methods of control of the threats a farmer can face during the production season. To match both objectives producing food of high quality in quantity, and guaranteeing a high level of human and environmental protection, a comprehensive set of regulatory texts has been developed and implemented since almost three decades, which imposes strict rules for the placing on the market and the use of chemicals in agriculture, including pesticides, at the European level. The Regulation (EC) No. 1107/2009, and previously the Directive No. 91/414 EEC, sets the requirements that every pesticide must meet before its registration and use, and that must allow to conduct a full risk assessment for users, consumers, and the environment. The Directive (EC) No. 2009/128 completes this legislation in establishing a framework for action to achieve a sustainable use of pesticides within the European Union, through actions that secure their use by farmers, including the training of users, the control of application machinery as well as the development of decision-making rules regarding the choice of the control methods by farmers so that to limit the use of chemicals to the minimum required and preferably in the context of Integrated Pest Management Programs.

This book aims at providing readers with an updated overview on the actions of different stakeholders to make the Sustainable Use of chemicals a reality.

The first chapter describes in detail the different regulatory texts constituting the legal framework for pesticides in Europe, and the rules regarding their implementation in the different countries of the European Union. It includes an analysis of their effectiveness at securing the catalogue of products, and illustrates through the example of risk management measures, the importance of coordination and harmonization within Europe to gather the greater number around common understanding of the tools developed and share experience for better, and quicker results.

The second chapter further emphasizes the critical importance of engaging all the stakeholders in actions that serves the four key aspects of a successful innovation in farming practices, namely governance, monitoring, management, and availability of collaborative tools. It proposes a methodology that was elaborated based on experiences, and a series of case studies are presented, where the methodology was applied and an integrative multiactor participatory network allowed to establish the level of mutual trust and collaboration needed for the network to a reach common objectives in the area of water management.

The third chapter focuses on certification systems and quality systems and contribute to create incentives for better practices and to bring trust in the system. The chapter successively addresses the role of internationally agreed standards and cross-country certification schemes in the trade of certified agricultural and food products, offers a comprehensive review of the regulatory framework in place for food quality and compliance with quality expectations, and discusses the benefits to farmers from economical to social, which contribute to sustainability of production systems in the short and long term.

The fourth chapter looks into the importance of the involvement of farmers in the development of their own practices toward more innovation so that to match the principle of the sustainable use directive. The analysis looks into the outcome of a number of European projects and summarizes the learnings regarding the success of the innovation intake among farmers, and suggests ways of improvement of innovation uptake through an observation of successful behaviors in collected experiences. The analysis also stresses the importance of communication, education, and training toward a responsible and sustainable management of chemicals in agriculture.

The last chapter then offers a series of examples of success stories in the implementation of the Sustainable Use directive, illustrating how some defined actions, tools, and collaborative work contribute to the building of the edifice. Bringing ambitious objectives to life, such as those agreed on and listed in the directive rely on the voluntary involvement of all the stakeholders involved, and necessitate to reach common understanding of the needs and implications, as well as consensus. Implementing these objectives takes time and resources, and although the current horizon scanning still highlights a number of research gaps and collaborative breakthroughs, it also unveils a number of success stories and a general agreement that reasoned and sustainable agriculture is the way forward.

ETTORE CAPRI[*], ANNE ALIX[†]
[*]Università Cattolica del Sacro Cuore,
Istituto di Chimica ed Ambiente, Piacenza, Italy
[†]Corteva Agrisciences™, the Agriculture Division of DowDupont™,
Abingdon, United Kingdom

# CHAPTER ONE

# Modern Agriculture in Europe and the Role of Pesticides

**Anne Alix\*, Ettore Capri[†,1]**

\*Corteva Agrisciences™, the Agriculture Division of DowDupont™, Abingdon, United Kingdom
[†]Università Cattolica del Sacro Cuore, Istituto di Chimica ed Ambientale, Piacenza, Italy
[1]Corresponding author: e-mail address: ettore.capri@unicatt.it

## Contents

1. Introduction    2
2. (EC) Regulation No. 1107/2009 on the Placing of Pesticides on the Market    4
   2.1 Requirements for Pesticides to be Placed on the Market    4
   2.2 Additional Measures to Further Reduce Exposure to Pesticides Through Additional Precautionary Phrases    6
3. Directive 2009/128 (EC): A Framework to Further Secure Pesticide Use and Application Toward Sustainability    11
   3.1 Improving the Awareness of Pesticide Users Through Training Programmes and Certification    11
   3.2 Improving the Quality and Efficacy of Pesticide Application Equipment    12
   3.3 Reducing the Use of Harmful Active Substances and Encourage the Development of Chemical and Nonchemical Alternatives    13
4. Effectiveness of the Regulatory Framework in the Context of Sustainable Use of Pesticides    14
   4.1 Effectiveness of (EC) Regulation 1107/2009: Screening the Pesticides to be Allowed on the Market for Safety    14
   4.2 Contribution of Directive 2009/138/EC to a Sustainable Use of Pesticides    16
5. Risk Management: Toward a Sustainable Agriculture at a Larger Scale    17
   5.1 Implementing the Risk Management Measures at Larger Scale    17
   5.2 Implementation of Precision Agriculture    18
6. Conclusions    20
References    21
Further Reading    22

## Abstract

The management of cropping systems is in constant evolution. Over the last decades, agriculture and crop management practices have shown deep changes, to allow the extension of cropped areas while ensuring an efficient control of pest populations and diseases. Plant protection products (PPP, also called pesticides) are part of these management practices, and the catalogue of products available to farmers has followed a parallel evolution to match farmers' expectations regarding efficacy and safety, but

*Advances in Chemical Pollution, Environmental Management and Protection*, Volume 2
ISSN 2468-9289
http://dx.doi.org/10.1016/bs.apmp.2018.04.001

© 2018 Elsevier Inc.
All rights reserved.

1

also the expectations of the public. The regulatory framework ruling the placing on the market of pesticides, as well as aspects of their use reflect these changes and shape the conditions of their use on a sustainable way. EC Regulation No. 1107/2009 bases the placing of pesticide products on the market on the demonstration that their use complies with defined protection goals guaranteeing a high level of safety for humans and the environment. Directive 2009/128/EC also called the "Sustainable Use Directive" extend the set of measures that, from the training and certification of users to the control of application machines and the development of effective alternative methods, should improve the safety level over the whole process.

This chapter offers an overview of the concept of sustainable use of pesticides, from the regulatory context with some insight on the changes that the implementation of the regulatory framework has already triggered. Examples are provided that illustrate the respective advantages of the two regulatory texts as well as the drawbacks they contain when considered separately. The chapter then reflects on current developments toward an ever safer agriculture meeting farmers' and general expectations as regards pesticides and their role in agriculture while matching human needs, including the development of an effective set risk management measures at the European level.

**Keywords:** Plant protection products, Pesticides, Regulation, Agriculture, Sustainable, Risk assessment, Risk management

# 1. INTRODUCTION

Agriculture is in constant evolution, and modern agriculture relates to a range of revolutionary steps that aimed at improving production, reducing workload and work penibility and since the last few decades actively improve human and environmental safety.

Alongside changes in practices at the producer's scale, this evolution has been echoed in the regulatory area, at the national level and further at the European level, and in the world. Particular attention was given to plant protection products (PPP), or pesticides, since the late 1950s at national level when the first registration procedure was set up in European countries, and then at European level at the end of the 1980s with the European Directive 91/414/EEC followed by the (EC) Regulation No. 1107/2009,[1] setting rules of the placing of PPP on the market. In Europe, the legislation has been extended to the development of a strategy for a "Sustainable Use of Pesticides."[2a]

What is meant by a "sustainable use" of pesticides is actually not defined in the legislation. The objectives are to "*reduce risks and impacts of pesticide use on human health and the environment and at encouraging the development and introduction of integrated pest management and of alternative approaches or techniques in order to reduce dependency on the use of pesticides."*[2a] An interpretation of this may be that the use of pesticides in the long term should be of limited *"footprint"* to human health and the environment and that any effective alternatives to pesticide applications are encouraged.

Two regulatory texts currently constitute the legislation on pesticides and should allow to reach this low *"footprint"* objective laid down in the principle of sustainability. First, pesticides are subject to a very comprehensive *risk assessment* for each active substance and for the products containing that substance before they can be authorized for use. This risk assessment is a prerequisite in (EC) Regulation No. 1107/2009 and aims at preventing risks at the source. As a consequence, granting an authorization for a PPP implies that it has been established that *"under normal conditions of use, the uses authorised exert no unacceptable effects on human and animal health, and on the environment."* Directive 2009/128/EC then completes this legislation in establishing a framework for action to achieve a sustainable use of pesticides within the European Union. This second piece of legislation notably addresses more specifically the use phase of pesticides, which is a key element for the determination of the overall risks that pesticides may pose. This framework includes the following key actions:
- improve the awareness of pesticide users (in particular professional users) by ensuring better training and education;
- improve the quality and efficacy of pesticide application equipment;
- implement controls of the application methods and of pesticide distribution;
- reduce the use of harmful active substances and encourage the development of chemical and nonchemical alternatives;
- develop a set of indicators aiming at reporting on progress made in the fulfilling of the objectives of the Directive.

To meet with these more general objectives Member States are invited to develop a "National Action Plan" that aim at setting altogether the individual measures listed in Directive 2009/128/EC. Both pieces of legislation are being implemented in Europe and the Member States and their efficacy and improving the safety of pesticide use can be appreciated through an analysis of the catalogue of products as well as through the first reports on the implementation of the Sustainable Use Directive published by the European Commission.[2b] Practical implementation of the legislation has also triggered the development of efficient risk management tools which, when aligned with broader pieces of legislation as, for example, the Common Agricultural Policy[3] and environmental regulatory framework, provide a meaningful and efficient toolbox to meet sustainability objectives agriculture today and in the future.

## 2. (EC) REGULATION NO. 1107/2009 ON THE PLACING OF PESTICIDES ON THE MARKET

### 2.1 Requirements for Pesticides to be Placed on the Market

(EC) Regulation No. 1107/2009/EC and previously Directive 91/414/EEC lay down the criteria for the placing of pesticides on the market.[1,4] The regulation lists the content of the dossier that must be submitted to fulfill the data requirements established by the legislation for the active substances as well as for the formulations containing these active substances. These data are needed for a scientific analysis that will determined whether or not the active substance meets the "approval criteria" defined by the regulation. These criteria are the following:

The dossier is subject to a *"peer review"* that consists in an in-depth evaluation, by scientific experts of each Member State and the European Food Safety Authority (EFSA), of all the studies. A risk assessment is undertaken for the uses of formulated product as recommended to the user. The conclusions of this evaluation are the basis for decision making. At the European level, it is decided whether or not the active substance may be included on a positive list, or "approved" under (EC) Regulation 1107/2009/EC. In Member States at national level, the evaluation pursues for each of the formulated products containing the active ingredient and similarly authorizations may be granted for the uses of these products that comply with decision-making criteria for acceptable risks. Each use of the product is evaluated so that it is thereafter defined to bring the minimal application rate at optimized application frequency for an effective protection of the crop(s) with no impact on human health and the environment.

As regards the safety components of the protection goals, as observed from the criteria listed earlier, decision-making criteria include both hazard-based criteria and risk-based criteria. Hazard-based criteria eliminate de facto active substances that could be cancerogen; mutagen, toxic for the reproduction or display endocrine disruptive properties, is a POP (Persistent Organic Product) or PBT (Persistent, Bioaccumulable, and Toxic substance) as defined in the Annex II of the regulation. Risk-based criteria then complete the evaluation with the requirement that no unacceptable effect is anticipate for any of the aspects listed in Box 1. Risk assessment processes have been developed by EFSA to ensure acceptable risks to humans, i.e., the operator, worker, bystander,

**BOX 1 Approval Criteria for Active Substances After Scientific Review of the Dossier as per Article 4 of (EC Regulation No. 1107/2209) on the Placing on the Market of Plant Protection Products. All Criteria Must Be Met for the Approval to Be Proposed**

*Article 4 (EC) Regulation No. 1107/2009*

*Approval criteria for active substances*

1. *An active substance shall be approved in accordance with Annex II if it may be expected, in the light of current scientific and technical knowledge, that, taking into account the approval criteria set out in points 2 and 3 of that Annex, plant protection products containing that active substance meet the requirements provided for in paragraphs 2 and 3.*

   *The assessment of the active substance shall first establish whether the approval criteria set out in points 3.6.2 to 3.6.4 and 3.7 of Annex II are satisfied. If these criteria are satisfied the assessment shall continue to establish whether the other approval criteria set out in points 2 and 3 of Annex II are satisfied.*

2. *The residues of the plant protection products, consequent on application consistent with good plant protection practice and having regard to realistic conditions of use, shall meet the following requirements:*

   (a) *they shall not have any harmful effects on human health, including that of vulnerable groups, or animal health, taking into account known cumulative and synergistic effects where the scientific methods accepted by the Authority to assess such effects are available, or on groundwater;*

   (b) *they shall not have any unacceptable effect on the environment. For residues which are of toxicological, ecotoxicological, environmental, or drinking water relevance, there shall be methods in general use for measuring them. Analytical standards shall be commonly available.*

3. *A plant protection product, consequent on application consistent with good plant protection practice and having regard to realistic conditions of use, shall meet the following requirements:*

   (a) *it shall be sufficiently effective;*

   (b) *it shall have no immediate or delayed harmful effect on human health, including that of vulnerable groups, or animal health, directly or through drinking water (taking into account substances resulting from water treatment), food, feed or air, or consequences in the workplace or through other indirect effects, taking into account known cumulative and synergistic effects where the scientific methods accepted by the Authority to assess such effects are available; or on groundwater;*

   (c) *it shall not have any unacceptable effects on plants or plant products;*

*Continued*

---

**BOX 1 Approval Criteria for Active Substances After Scientific Review of the Dossier as per Article 4 of (EC Regulation No. 1107/2209) on the Placing on the Market of Plant Protection Products. All Criteria Must Be Met for the Approval to Be Proposed—cont'd**

(d) *it shall not cause unnecessary suffering and pain to vertebrates to be controlled;*

(e) *it shall have no unacceptable effects on the environment, having particular regard to the following considerations where the scientific methods accepted by the Authority to assess such effects are available:*

  (i) *its fate and distribution in the environment, particularly contamination of surface waters, including estuarine and coastal waters, groundwater, air, and soil taking into account locations distant from its use following long-range environmental transportation;*

  (ii) *its impact on nontarget species, including on the ongoing behavior of those species;*

  (iii) *its impact on biodiversity and the ecosystem.*

---

consumers, and the environment, i.e., water quality (surface- and groundwater), to organisms of the aquatic and terrestrial ecosystems. This includes birds and mammals of agricultural ecosystems, invertebrate fauna including nontarget flora, soil micro- and macroorganisms in the field, and aquatic organisms of border water bodies.

## 2.2 Additional Measures to Further Reduce Exposure to Pesticides Through Additional Precautionary Phrases

In addition to the approval principles, (EC) Regulation 1107/2009 plan the possibility of risk-mitigation measures accompany the product, which aim at describing specific conditions of application of the product that will further limit exposure and transfers of the active substance and its residues[12] toward environmental compartment (Table 1). These risk-mitigation measures are most often derived from the risk assessment, but may also be generic. These phrases may also be generic, as, for example, the following condition of application: "*SP1: Do not contaminate water with the product or its container (Do not clean application equipment near surface water/Avoid contamination via drains from farmyards and roads),*" which is of generic nature as it relates to good practice. On the other hand, the phrase: "*SPe2: To protect groundwater/aquatic organisms*

**Table 1** Standard Phrases for Safety Precautions for the Protection of Humans or the Environment as Referred to in Article 16 of Directive 91/414/EC (EC, 2003)[4]

| Protection Goal | No | Standard Phrase | Condition for Attribution |
|---|---|---|---|
| Surface water | SP 1 | Do not contaminate water with the product or its container (do not clean application equipment near surface water/avoid contamination via drains from farmyards and roads) | All plant-protection products should be labeled with this following phrase, which should be supplemented by the text in parentheses, as appropriate |
| Operators | SPo1 | After contact with skin, first remove product with a dry cloth and then wash the skin with plenty of water | The phrase shall be assigned for plant-protection products containing ingredients which may react violently with water, such as cyanide salts or aluminium phosphide |
| | SPo2 | Wash all protective clothing after use | The phrase is recommended when protective clothing is required to protect operators. It is obligatory for all plant-protection products classified as T or T+ |
| | SPo3 | After igniting the product, do not inhale smoke and leave the treated area immediately | The phrase may be assigned to plant-protection products used for fumigation in cases where the use of a respiratory mask is not warranted |
| | SPo4 | The container must be opened outdoors and in dry conditions | The phrase should be assigned to plant-protection products containing active substances which may react violently with water or damp air, such as aluminium phosphide, or which may cause spontaneous combustion, such as (alkylenebis–) dithiocarbamates. This phrase may also be assigned to volatile products classified with R20, 23, or 26. Expert judgment must be applied in individual cases, to assess whether the properties of the preparation and the packaging are such as to cause harm to the operator |
| | SPo5 | Ventilate-treated areas/greenhouses thoroughly/time to be specified/until spray has dried before reentry | The phrase may be assigned to plant-protection products used in greenhouses or other confined spaces, such as stores |

*Continued*

**Table 1** Standard Phrases for Safety Precautions for the Protection of Humans or the Environment as Referred to in Article 16 of Directive 91/414/EC (EC, 2003)[4]—cont'd

| Protection Goal | No | Standard Phrase | Condition for Attribution |
|---|---|---|---|
| Ground water/soil | SPe1 | To protect groundwater/soil organisms do not apply this or any other product containing (identify active substance or class of substances, as appropriate) more than (time period or frequency to be specified) | The phrase shall be assigned to plant-protection products for which an evaluation according to the uniform principles shows for one or more of the labeled uses that risk-mitigation measures are necessary to avoid accumulation in soil, effects on earthworms or other soil-dwelling organisms or soil microflora and/or contamination of groundwater |
| Ground water/aquatic organisms | SPe2 | To protect groundwater/aquatic organisms do not apply to (soil type or situation to be specified) soils | The phrase may be assigned as a risk-mitigation measure to avoid any potential contamination of groundwater or surface water under vulnerable conditions (e.g., associated to soil type, topography, or for drained soils), if an evaluation according to the uniform principles shows for one or more of the labeled uses that risk-mitigation measures are necessary to avoid unacceptable effects |
| Aquatic organisms/nontarget plants and arthropods | SPe3 | To protect aquatic organisms/nontarget plants/nontarget arthropods/insects respect an unsprayed buffer zone of (distance to be specified) to nonagricultural land/surface water bodies | The phrase shall be assigned to protect nontarget plants, nontarget arthropods, and/or aquatic organisms, if an evaluation according to the uniform principles shows for one or more of the labeled uses that risk-mitigation measures are necessary to avoid unacceptable effects |
| Aquatic organisms/nontarget plants | SPe4 | To protect aquatic organisms/nontarget plants do not apply on impermeable surfaces such as asphalt, concrete, cobblestones, railway tracks, and other situations with a high risk of runoff | Depending on the use pattern of the plant-protection product, Member States may assign the phrase to mitigate the risk of runoff in order to protect aquatic organisms or nontarget plants |

| | | | |
|---|---|---|---|
| Birds/wild mammals | SPe5 | To protect birds/wild mammals the product must be entirely incorporated in the soil; ensure that the product is also fully incorporated at the end of rows | The phrase shall be assigned to plant–protection products, such as granules or pellets, which must be incorporated to protect birds or wild mammals |
| Birds/wild mammals | SPe6 | To protect birds/wild mammals remove spillages | The phrase shall be assigned to plant–protection products, such as granules or pellets, to avoid uptake by birds or wild mammals. It is recommended for all solid formulations, which are used undiluted |
| Birds | SPe7 | Do not apply during the bird breeding period | The phrase shall be assigned when an evaluation according to the uniform principles shows that for one or more of the labeled uses such a mitigation measure is necessary |
| Bees | SPe8 | Dangerous to bees. To protect bees and other pollinating insects do not apply to crop plants when in flower. Do not use where bees are actively foraging. Remove or cover beehives during application and for (state time) after treatment. Do not apply when flowering weeds are present. Remove weeds before flowering. Do not apply before (state time) | The phrase shall be assigned to plant–protection products for which an evaluation according to the uniform principles shows for one or more of the labeled uses that risk-mitigation measures must be applied to protect bees or other pollinating insects. Depending on the use pattern of the plant–protection product, and other relevant national regulatory provisions, Member States may select the appropriate phrasing to mitigate the risk to bees and other pollinating insects and their brood |
| Good agricultural practice | SPa1 | To avoid the buildup of resistance do not apply this or any other product containing (identify active substance or class of substances, as appropriate) more than (number of applications or time period to be specified) | The phrase shall be assigned when such a restriction appears necessary to limit the risk of development of resistance |

*Continued*

**Table 1** Standard Phrases for Safety Precautions for the Protection of Humans or the Environment as Referred to in Article 16 of Directive 91/414/EC (EC, 2003)[4]—cont'd

| Protection Goal | No | Standard Phrase | Condition for Attribution |
|---|---|---|---|
| Side effects from rodenticides | SPr1 | The baits must be securely deposited in a way so as to minimize the risk of consumption by other animals. Secure bait blocks so that they cannot be dragged away by rodents | To ensure compliance of operators the phrase should appear prominently on the label, so that misuse is excluded as far as possible |
| Side effects from rodenticides | SPr2 | Treatment area must be marked during the treatment period. The danger from being poisoned (primary or secondary) by the anticoagulant and the antidote against it should be mentioned | The phrase should appear prominently on the label, so that accidental poisoning is excluded as far as possible |
| Side effects from rodenticides | SPr3 | Dead rodents must be removed from the treatment area each day during treatment. Do not place in refuse bins or on rubbish tips | To avoid secondary poisoning of animals the phrase shall be assigned to all rodenticides containing anticoagulants as active substances |

*do not apply to (soil type or situation to be specified) soils*" is to be appear on the label when the risk assessment identified a specific vulnerability condition to be accounted for by users. These measures and related phrases are reproduced on the product's label.

A complete set of precautionary, or mitigation measures, has been developed for further protection of operators (SPo) and the environment (SPe), for which they propose risk management measures designed for the protection of groundwater (SPe1 and 2), organisms and ecosystems boarding treated areas like water bodies and vegetated strips including the invertebrate fauna (SPe3), surface water from transfers through runoff (Spe4), granivorous birds and mammals (SPe5 and 6), bird reproduction (SPe7), and pollinators (SPe8). These are reproduced in Table 1.

## 3. DIRECTIVE 2009/128 (EC): A FRAMEWORK TO FURTHER SECURE PESTICIDE USE AND APPLICATION TOWARD SUSTAINABILITY

Directive 2009/128/EC completes (EC) Regulation No. 1107/2009 through a list of complementary actions at the farm scale and beyond, aiming at further securing the use of pesticide products but also contributing to reinforce the reasoned, rational rather than prophylactic use of chemical control methods over nonchemical alternatives.

### 3.1 Improving the Awareness of Pesticide Users Through Training Programmes and Certification

Awareness and understanding of the implication of labeling instructions is critical factor to ensure that products are applied according to the directions of the labeling and therefore in respect of conditions being safe for both the worker and the environment. These conditions have been designed so that to ensure that safety and protection goals are met and they therefore deserve the attention of users so that protection goals are actually met.

The European Commission reports a high level of compliance in the area of training and certification for professional users, but also distributors and advisors, within the European Member States,[2b] which is remarkable having in mind the diversity of statuses of distributors and advisors between countries. The compliance to the training may be checked through official controls and is reported as high as 95% in the countries where data are recorded.

Another measure is the restriction on the sale of some pesticides to nonprofessional users. This measure has been implemented in the 28 European countries according to the Commission report.[2b] The measure can, for example, limit access of nonprofessionals to the sole products that can be applied with standard personal protections or do not display toxic properties as from the label.

Industry may participate to ad hoc training in the case of specific applicator devices, for example, to be used by trained professional only such as injectors, and this enters into dedicated the stewardship plans that accompany the registration of a product.

## 3.2  Improving the Quality and Efficacy of Pesticide Application Equipment

Another set of measures listed in Directive 2009/128/EC aim at improving the quality and efficacy of pesticide application equipment, as through the implementation of technical control of application methods, mainly sprayers. The technical control of sprayers is included in 26 of the 28 NAP and was preexisting the implementation of the Directive.[2b] Usually sprayers enter the market after having been checked for compliance with the manufacturer specifications. Additional controls take place at regular intervals, and in some countries controls have been set by professional organizations. Sprayers subject to control are identified on the basis of the date at which they were purchased. Harmonization work is also an important factor to support Member States in the implementation of these measures, and a European working group is developing harmonized approach for the inspection of sprayers that include common interpretation of the regulation, identification, and development of the ISO norms to include in the controls and training of the inspection services (http://spise.julius-kuehn.de/dokumente/upload/SPISE_ADVICE_1-2017.pdf).

Spray drift reducing nozzles are often recommended as a measure to further reduce spray drift during pesticide application. This measure is implemented in 15 Member States and may appear on the national label where requested.[5a] These nozzles are developed in compliance with technical specification so that they allow a drift reduction that ranges, depending on the models, between 50% and 90%. As for sprayers controls allow to check the compliance to the specification, which may be mutually recognized between countries. As above a dedicated working group is working at facilitating data generation, analysis, and data sharing and offers a website to make the information generated accessible to the stakeholders in the Member States (https://www.spraydriftmitigation.info/).

Finally, for products being applied as seed coating and therefore together with the seed, specific technical control has been developed for drillers. Drillers are not explicitly mentioned in Directive 2009/128/EC. Some drillers may generate dusts during sowing, and these dusts may contain significant amounts of product residues. This is the case of planters that use airflow to push seeds into the machine and reject the air upward. Several countries have already taken this issue into account in implementing risk management measures that allow to limit the level of dust production and dispersion, among which the equipment of drillers with drift reducing devices (deflectors) and the restriction of drilling to limited wind conditions.[5b–7]

## 3.3 Reducing the Use of Harmful Active Substances and Encourage the Development of Chemical and Nonchemical Alternatives

Directive 2009/128/EC also promotes the development of alternative methods with the aim to broaden the range of crop protection solutions, and to reduce "pesticide use for the substances of concern."[2] The last issue is also to be linked to Regulation 1107/2009/EC, and to the approval criteria relative to the substances of concern as defined in the cases of substances being "candidate for substitution" for which alternatives are explicitly discussed in the context of the regulation.

Identifying effective alternative methods remains difficult, as illustrate by the number of emergency authorizations that are still being delivered in Member States for pesticides in order to extend their use to a specific crop or pest where a regular authorization is not available and this to prevent crop losses. When available, alternative measures also have to be brought to farmers' knowledge, through, for example, guides or bulletins. To be accepted and implemented by farmers, alternative control methods must provide satisfying and reliable level of pest, weed, or disease protection, at equivalent costs or lower than conventional methods. Additional considerations relative to easiness of implementation are also important. Often alternatives prove to be effective for protected crops, but their use in open field crops requires more developments to achieve reproducible results.

The Directive recommends that all the alternatives are considered by the farmer in preparing his seasonal programme, and recommends that the use of pesticides enters in broader Integrates Pest Management Programmes (IPM). These programmes are designed around low-pesticide input and include a range of practices aiming at reducing the vulnerability of the crop to weeds, pests, and diseases, and limit the use of pesticides to levels that are economically and ecologically justified. In Member States, tools to predict and diagnose the presence of crop threats are now well implemented as a prerequisite to IPM implementation; however, the diversity of tools actually available together with difference in agricultural practices, climate, and crop dynamics varies between countries thus leading to different definition of the criteria and objectives.[2b] (EC) Regulation 1107/2009 also promotes the development of low-risk products, which can contribute toward reasoned and integrated pest management objectives.[1]

## 4. EFFECTIVENESS OF THE REGULATORY FRAMEWORK IN THE CONTEXT OF SUSTAINABLE USE OF PESTICIDES

### 4.1 Effectiveness of (EC) Regulation 1107/2009: Screening the Pesticides to be Allowed on the Market for Safety

The regulatory process implemented in the early 1990s at the European level was initiated on ca 1200 active substances that were in use in the European countries. The Directive requested the owners of these substances to provide a regulatory dossier responding to strict data requirements addressing chemical and physical properties, toxicological effects, effects on nontarget species of the environment, and their fate in the different compartments of the environment after use (soil, water, air, and plants). Incomplete dossiers lead to a phase out of the active substance and of the pesticide formulations in which they entered. Complete dossiers were allowed to enter the evaluation process, during which the risks to human, including users, consumers, and bystanders and to the environment, were evaluated for each use. Following the first review of the active substances catalogue, ca 350 were included on the positive list at the European level and Member States revised their national registrations accordingly. Newly developed active substances follow the same process and must respond to the same criteria. The approval is granted for 10 years and the European process plans a renewal procedure that reviews the content of the dossier on a regular basis so that to ensure that the scientific risk assessment is always aligned with advances in regulatory science and that any unforeseen risk is taken into account. (EC) Regulation 1107/2009 that updated Directive 91/414 (EC) follows the same principles and introduced additional considerations as regards low-risk substances.

Industry actively participates to the safety of pesticide developments, in integrating early in the discovery process a number of testing that are included in the data requirements set by the Regulation. This way, molecules are screened for safety before to get to the end of the selection process and technically, new pesticides are all screened to display no concerning effects on human health, limited persistence in the environment or side effects on nontarget organisms, and a quick loss of the toxophore leading to a degradation process containing only nontoxic metabolites. The achievement of limited toxicity while displaying a satisfying efficacy on the target relies on specific mode of actions, behavior in the plant after treatment, and/or specific metabolism in target vs nontarget organisms. The increased selectivity to the target has increased the safety of the products being now in use and proposed as new active substances. The drawback of selectivity is in the increasing number of pests, diseases, or weeds that are not covered anymore by chemical control spectrum and that now rely on alternative methods to be controlled.

The development of the labeling has also contributed to further secure pesticide use. The advantages of product-specific labeling measures are those of a tailor-made system that recommends measures being adapted to the product and thus meaningful to the user. Recommendations and directions for use lose their meaning if recommended generically as users who doubt on the need for extraprecautionary measures for some products may then underestimate the level of precautions to be taken in general and adapt their behavior. This is why it is important to adapt the level of precautions to the product properties itself so that to maintain the level of awareness and alert in end users.

The labeling system has, however, some drawbacks that in relation to the level of detail given in the instructions. It has been observed that reading through a long list of sentences on the label may discourage farmers to implement them all. In addition, a frequent feedback from farmers is products are actually often used at rates lower than recommended on the label, this because they often enter in a treatment program that comprises other components, or because the rate may be adapted to level of infestations, for example. It is thus often wondered how risk management recommendations should be interpreted in the light of crop management practice. All these issues indicate the need for, a minima, an appropriate communication to farmers on how these recommendations should be implemented at the scale of the farm. User-friendly applications that may be downloaded on tablets or phones may also provide farmers with simple tools to aid the application step and facilitate the integration of a treatment in a broader programme, including in IPM programmes.

A step forward may also rely in a further evolution of the agricultural practices that would include default risk-mitigation options. As an example, in France all products registered include a default buffer zone to a water body of 5 m.[8] The systematic implementation of such a mitigation measure can significantly contribute to simplify labeling and the actual implementation of the measure itself, hence making of it a good agricultural practice.

## 4.2 Contribution of Directive 2009/138/EC to a Sustainable Use of Pesticides

In extending the set of measures to the conditions of use of products and through the promotion of application methods that would contribute to limit the footprint of pesticides in the agricultural area, Directive 2009/128/EC completes the authorization system that is built to guarantee a safe use of each product considered individually. Indeed the effectiveness of measures being recommended on the labeling may be compromised by a lack of awareness of users of the importance of complying application recommendations or by the use of inappropriate application material. Thus, the Directive, in targeting the whole process from the decision to treat to the application itself, makes that the work undertaken since about 25 years in the frame of pesticide security may be even more effective.

As stated earlier, a number of objectives of the Directive are being implemented in the European Member States.[2b] This implementation, however, occurs at national pace and is not necessarily tracked in a way that provides enough visibility to the Commission nor to the stakeholders. This lack of visibility concurs to the general perception that the level of implementation is insufficient, and often calls for more "clear cut" or "blanket" measures, that provide quick visibility without effectively addressing the problem. Discussions on the development of "risk indicators" to measure the effectiveness of national action plans are a good illustration of this debate. A number of countries have hence worked on the definition of a risk indicator that could be used to reflect yearly progress on the evolution of practices toward a more sustainable use of pesticides. The preferred models so far rely on the assumption that the overall footprint of pesticides is proportionate to their application rate. Such indicators are easy to develop since they are based on data on application rates, market figures, and/or cultivated surfaces, which are easy to generate. Such indicators cannot reflect the reasoning behind the use not the approval, the relative positioning of a product vs alternatives or in a treatment programme. Work in this direction has been conducted in some cropping systems where "Ecosystem Services" based approaches were used as a support to decision in designing plant protection programmes in conventional agriculture or in IPM.[9,10] These approaches were developed at the production scale but may be adapted to derive toolsets to help design policy-making processes and inform the development of these indices.

Another drawback relies in that risk–mitigation measures, under the (EC) Regulation 1107/2009 or the Directive, was being developed in each Member State, based on existing risk management tools, development, and policies. While this allowed for a quick implementation and effectiveness at reducing pesticide unintentional effects, this proved to be counterproductive relative to the efforts development to increase pesticide safety through harmonization of their evaluation and management at European level.

Initiatives developed in this context as, for example, in 2009 the "think tank" OPERA that worked at using and communicating on the potential of existing scientific researches and knowledge to support stakeholders in their decisions concerning risk management related to pesticides and the environment.[11] In 2011, a European initiative conducted to the development of a risk management toolbox dedicated to pesticides used in agriculture, which is further presented later.

## 5. RISK MANAGEMENT: TOWARD A SUSTAINABLE AGRICULTURE AT A LARGER SCALE

### 5.1 Implementing the Risk Management Measures at Larger Scale

As explained earlier, environmental risk–mitigation measures are a key component in defining the conditions of use of pesticides in crop protection.[1,12] These risk–mitigation measures derive directly from the evaluation of pesticide products and the risk assessment conducted for each use, are reported in the approval regulations for an implementation in European Member States[13] and at national level, and are reported on the labeling in the form of Safety Precaution Phrases, according to Regulation (EU) No 547/2011.[12]

The need for a harmonized set of risk–mitigation measures has been identified a few years ago and lead to a European initiative, called MAgPIE, with the objective was to develop a toolbox of risk–mitigation measures designed for the use of pesticides for agricultural purposes, and thus contribute to a better harmonization of their development and use within Europe. Two workshops were organized that gathered risk assessors and risk managers of 24 European countries, industry, academia, and agronomical advisors/extension services. Experts worked on the basis of an inventory of the environmental risk-mitigation measures in use in European countries, collected via questionnaires. Measures addressed groundwater protection, surface water protection, and the protection of off-crop /off-field areas and in-crop/ in-field area. The risk-mitigation tools inventoried were ranked based on the following criteria:

- Implementation/advancement level: from well-implemented tools in countries to tools on which insufficient knowledge or confidence were available;
- Regulatory aspects: regulatory status of the tool, from the straight implementation of a legislation in place to simple good farming practices, possible regulatory hurdles associated to a tool as well as options to resolve them;
- Economical aspects (costs);
- Possibility to measure the efficacy of the tool;
- Possibility to relate to the risk assessment, i.e., to develop a risk assessment that accounts for the risk-mitigation tool quantitatively or qualitatively.

On the basis of this analysis, recommendations were made with regards to the tools ready to be used in most Member States, and dedicated risk-mitigation technical sheets were prepared to that to describe in detail the methodology of their implementation and management in the field. The toolbox was then used to draft a set of revised the Safety Precaution Phrases as per in Regulation (EU) No 547/2011, that would support the implementation of the risk-mitigation measures listed in the toolbox, for further consideration by the European Commission and Member States. The proceedings of the workshop are published[5a] and are under implementation in some Member States. The benefit of such approaches is its reliance on practices and therefore contains measures that are de facto acceptable by users: in collecting experiences in the field where practices are experienced and evolve, and build a set of tools that proved to be useful to farmers so that they may be shared, and further developed, by others. The implementation at national level is facilitated by the consistency and cross-references with policies already in place such as the Sustainable Use Directive with the actions on training and the CAP,[3] through a common use of farmland features such as field margins or noncultivated land.

## 5.2 Implementation of Precision Agriculture

Precision agriculture (PA) gathers a range of application techniques aiming optimizing crop production and crop protection to obtain the best results with targeted practices, and within these targeted product applications.

PA is part of modern farming management. It consists in observing, measuring and responding based on these measurements, to the "needs" with increased precision, so that the crop, or even the plant, gets exactly what it needs.

With regards to pesticides and fertilizers, it involves tools that can measure the needs, take into account intra- and intervariability between crops and adjust the amount of product delivered to the crop, or to the plant, to meet the needs. It comprises:

- In-row treatments and band applications, through application devices that can target the spray or application delivery (e.g., granule) in the row, instead of over the surface of the crop;
- Sensor-equipped devices (sprayers or drones) that detect the presence of weeds in the crop and target the spray onto the weeds;
- Diverse techniques of spray drift reduction and targeted spray, a review of which is available in Alix et al.[5a];
- Use of electronic system to read bar-code-labels of pesticide containers by the application device itself, so that the application directly interprets the label information, and proceeds to treatments automatically, so that to reduce the risk of misinterpretation and human errors and increases precision in the whole process;
- These electronic systems can also be connected to GIS data so that applications take into account geographical and landscape specificities such as vulnerability of the land to runoff or leaching for example, when these data are available and mapped.

As an example from Germany the research and development project "Pesticide Application Manager (PAM)-Decision support in crop protection based on terrain-, machine-, business - and public data" is working on these approaches and details may be found in Golla et al.[14] for a description of the founding principles and Alix et al.[5a] for a dedicated risk mitigation technical leaflet). This consortium develops tools to automate and optimize the processes for planning, implementing and documenting pesticide application. The overall aim is to make pesticide application less error-prone.

Implementation is still in development in most countries as it requires field mapping of weeds, disease or pest occurrences, or off- and in-field vulnerabilities translated into Global Positioning System (GPS) format as input to in-cab GPS connected to spray machinery to ensure delivery of product to desired targets. It requires investment in GPS devices and associated linkage with machinery and a degree of technical sophistication.

In addition, the costs to farmers to upgrade their equipment needs to be taken into account in deciding upon implementation strategies and related incentives. However expected benefits are significant, as these strategies will allow for product delivery tailored to the needs, taking into account agroenvironmental data as well as weather conditions, hence reduced costs to farmers. These strategies also allow automated data recording and storage, to further feed and improve the technique itself and the global data system.

Mapped information on wildlife presence when available can also be used in order to reduce their exposure. As an example, the exposure of birds can be reduced during the breeding period, in limiting pesticide application to patches in the field. This is particularly applicable to herbicides, as weeds in fields can occur in patches, as a result of soil variation, general variability in plant distributions and as a result of previous weed management operations. Patch spraying provides an opportunity for two contrasting biodiversity enhancing approaches. If fields contain discrete patches of noxious unwanted species, the technology can be used to focus application to these patches. Spatially selective weed management has considerable potential for allowing managing populations of species to be protected to achieve a balance between production and conservation objectives. The main barriers to wider uptake at present are the capital costs of the equipment and the time involved in producing weed maps. Weeds for which treatments are required should grow large enough to be readily identified, without prejudicing control if required, and allow for producing maps. The use of spatially explicit methods may also be used for selectively restricting spaying in ecological hotspots such as nesting sites, especially for sensitive species. Successful implementation is reported in the UK, in collaboration with the RSPB.[15]

Overall, Precision Agriculture or Precision Farming is still under development due to the level of technology, data generation and gathering it involves and reports of routine implementation are still very limited.[2b] These techniques are however extremely promising and constitute a significant turn in tomorrow's agriculture and number of projects are being initiated to accelerate their development.

## 6. CONCLUSIONS

Both (EC) Regulation No. 1107/2009 and Directive 2009/128/EC contribute to provide farmers with a toolbox of products that meet a "high level of safety" to humans and the environment, as actually requested in both regulatory texts. The regulation provides a set of criteria that all products must meet to access the market and ensure that the products registered and in use match these protection goals. Risk management tools that are recommended in this context are product-designed and effective. Directive 2009/128/EC completes this system in giving a frame for the risk management measures over the whole process, from the application to the choice of the control methods that can enter into farmers' treatment programmes. Both regulations have significantly contributed to securing the use of pesticides in Europe and to meet the objective of a modern pesticide toolbox for purpose of modern agriculture demand.

An effective implementation of the measures recommended needs, however, to be connected to field situations and acknowledged within the European area. The challenge for stakeholders and more particularly for decision makers relies in getting a common understanding and shared visions on the implication of both pieces of legislation for the agricultural systems. A step forward relies in the development of common sets of methods, indicators, and tools, as, for example, the toolbox developed for risk management purposes, built on the basis of interconnected research outcomes and on the mutual recognition of the most effective tools. Precision farming and precision agriculture, in using tailored and targeted applications where and when needed, can provide additional level of control and reproducibility of applications. Ultimately, this will set the basis of Safe Agricultural Practices, pieces of which are being implemented already in some areas awaiting to be shared, understood, and applied.

## REFERENCES

1. EC. *Regulation (EC) No 1107/2009 of the European Parliament and of the council of 21 October 2009 concerning the placing of plant protection products on the market and repealing council directives 79/117/EEC and 91/414/EEC.* OJ L 309, 24.11.200920091.
2a. EC. *Directive 2009/128/EC of the European Parliament and of the council of 21 October 2009, establishing a framework for community action to achieve the sustainable use of pesticides.* OJ L 309, 24.11.2009200971.
2b. EC. *Overview report on the Sustainable Use of Pesticides.* 2017. ISBN: 978-92-79-52987-0, https://doi.org/10.2875/846869; http://ec.europa.eu/food/audits-analysis/overview_reports/details.cfm?rep_id=114.
3. EC. *European Commission, "CAP reform—an explanation of the main elements (MEMO/13/621),"* Brussels: European Commission; 2013.
4. EC. DIRECTIVE 91/414/EEC, council directive of 15 July 1991 concerning the placing of plant protection products on the market (91/414/EEC). *Off J Eur Union* 2010. L 0414: 01.08.2010.
5a. Alix A, Brown C, Capri E, Goerlitz G, Golla B, Knauer K, et al. *Mitigating the risks of plant protection products in the environment.* SETAC editions; 2017. available online, https://www.setac.org/magpie.

5b. JORF. Arrêté du 13 avril 2010 modifiant l'arrêté du 13 janvier 2009 relatif aux conditions d'enrobage et d'utilisation des semences traitées par des produits mentionnés à l'article L. 253-1 du code rural en vue de limiter l'émission des poussières lors du procédé de traitement en usine. *NOR: AGRG1007789A*. 2018. Version consolidée au 02 mai 2018.

6. AGES. http://www.ages.at/LandwirtschaftlicheSachgebiete/Bienen/MaismittelundBienen/; 2010.

7. BVL. www.bvl.bund.de/DE/08_PresseInfothek/01_FuerJournalisten/01_Presse_und_Hintergrundinformationen/01_PI_und_HGI/PSM/2011/2011_07_08_hi_neonikotinoide.html?nn=1486928; 2011.

8. JORF. *Arrêté du 12 septembre 2006 relatif à la mise sur le marché et à l'utilisation des produits visés à l'article L. 253-1 du code rural et de la pêche maritime NOR: AGRG0601345A*. 2006. ersion consolidée au 04 mars 2015.

9. Deacon S, Alix A, Knowles S, Wheeler J, Tescari E, Alvarez L, et al. Integrating ecosystem services into crop protection and pest management: case study with the soil fumigant 1,3-D and its use in tomato production in Italy. *Integr Environ Assess Manag* 2016;**12**(4):801–10. https://doi.org/10.1002/ieam.1761.

10. Holt A, Alix A, Thompson A, Maltby L. *Food production, ecosystem services and biodiversity: we can't have it all everywhere*. *Sci Total Environ* 2016;**573**:1422–9. 15 December 2016, https://doi.org/10.1016/j.scitotenv.2016.07.139.

11. OPERA. http://www.opera-indicators.eu/eng/background/the-directive-on-the-sustainable-use-of-pesticides-1.html; 2009.

12. EC. *Commission regulation (EU) No 547/2011 of 8 June 2011 implementing regulation (EC) No 1107/2009 of the European Parliament and of the council as regards labelling requirements for plant protection products*. OJ L 155, 11.06.2011, 2011176.

13. EC. *Commission implementing regulation (EU) No 540/2011 of 25 May 2011 implementing regulation (EC) No 1107/2009 of the European Parliament and of the council as regards the list of approved active substances*. 2015. updated 01.01.2015.

14. Golla B, Enzian S, Gutsche V. GIS-aided approaches in considering local and regional landscape conditions in the pesticide use regulations process. In: Rossing WAH, Poehling HM, Burgio G, editors. *Landscape management for functional biodiversity*. IOBC WPRS Bulletin; 2003. p. 59–64. ISSN/ISBN: 92-9067-152-X.

15. Natural England. *Entry level stewardship - environmental stewardship handbook*. 4th ed. 2013. Natural England, www.naturalengland.org.uk/es.

## FURTHER READING

16. EC. *Council directive of 15 July 1991 concerning the placing of plant protection products on the market (91/414/EEC)*. 1991. OJ L 230/1. 19.08.1991.

CHAPTER TWO

# The Multiactor Approach Enabling Engagement of Actors in Sustainable Use of Chemicals in Agriculture

Els Belmans*, Paul Campling†, Elien Dupon‡, Ingeborg Joris†,
Eva Kerselaers*, Saskia Lammens§, Lies Messely*, Ellen Pauwelyn‡,
Piet Seuntjens†,¶,‖,1, Erwin Wauters*

*Flanders Research Institute for Agriculture, Fisheries and Food, Social Sciences Unit, Merelbeke, Belgium
†Flemish Institute for Technological Research, VITO, Environmental Modeling Unit, Mol, Belgium
‡Inagro, Rumbeke-Beitem, Belgium
§Flanders Environment Agency, Buitendienst IJzer, Leie en Brugse Polders, Oostende, Belgium
¶Ghent University, Ghent, Belgium
‖University of Antwerp, Institute of Environment and Sustainable Development, Wilrijk, Belgium
1Corresponding author: e-mail address: piet.seuntjens@vito.be

## Contents

1. Introduction                                                                24
2. General Methodology                                                         25
3. The Multiactor Approach                                                     27
    3.1 What Is a Multiactor Approach?                                          27
    3.2 Why a Multiactor Approach?                                             27
    3.3 Planning of the Multiactor Approach in Action Labs                     28
    3.4 Guidance for the Implementation of a Multiactor Approach               32
4. Water Governance                                                            34
    4.1 General                                                                34
    4.2 Water Governance Framework for Water Quality and Agriculture           35
    4.3 Overview of the Governance Framework in Three Steps                    37
5. Participatory Monitoring                                                    49
6. Best Management Practices                                                   50
7. Collaborative Software Tools                                                50
8. Results From Example Case Studies                                          51
    8.1 Cicindria Case Study                                                   51
    8.2 Bollaertbeek Case Study                                                52
    8.3 Vittel Case Study                                                      56
9. Conclusions                                                                 59
Acknowledgments                                                                59
References                                                                     59

*Advances in Chemical Pollution, Environmental Management and Protection*, Volume 2
ISSN 2468-9289
http://dx.doi.org/10.1016/bs.apmp.2018.03.001

© 2018 Elsevier Inc.
All rights reserved.

## Abstract

Water quality in rural areas is largely depending on farming practices. Despite numerous efforts to reduce nutrient and pesticide concentrations in surface water and ground-water and regulation (Nitrates Directive (91/676), Water Framework Directive (2000/60), Groundwater Daughter Directive (2006/118), Directive on Environmental Quality Standards (2008/105), Sustainable Use Directive (2009/128)), many water bodies are not in good status for nitrates and hot spots of contamination for phosphorous and pesticides persist. The stagnation of the water quality is not only due to a long-term storage of the agrochemicals in soil and groundwater systems but also to the lack of implementation of good agricultural practices and mitigation measures that prevent the chemicals to enter the water system. Over the past two to three decades, our under-standing of the functioning of the water system and the effect of farming practices and mitigation measures has increased tremendously, but somehow we fail to convert that knowledge into actual implementation at a scale which is necessary to create real improvement of water quality. We think that a paradigm shift from purely top-down regulation and enforcement to local actor engagement is needed to revert the trend. Therefore, we present here a multiactor approach to engage actors at the scale of water system catchment (river or aquifer) or a so-called action lab. Within an action lab, key actors are involved in setting up new governance strategies including alternative financing regimes, participatory monitoring approaches, best management practices, and collaborative software applications to facilitate the process of implementation of mitigation measures. This chapter describes methodological approaches to create engagement in a multiactor setting, including the four key aspects on governance, monitoring, management and collaborative tools, and a few example stories related to water quality.

**Keywords:** Water quality, Agriculture, Agrochemicals, Rural, Nutrients, Pesticides, Multiactor, Governance, Monitoring, Mitigation, Management practices

## 1. INTRODUCTION

High-quality, safe, and sufficient drinking water is essential for life: we use it for drinking, food preparation, and cleaning. However, more than half of the river and lake water bodies in Europe are reported to be in less than good ecological status and about 25% of groundwater across Europe is in poor chemical status.[1] The increased uptake of measures to reduce nitrogen pollution from agriculture and improvements in waste water treatment have led to a steady reduction in average nitrogen concentrations in rivers from 2.7 to 2.1 mg $NO_3$/l (1992–2012), and average concentrations in ground-water are well below the Groundwater Quality Standard of 50 mg $NO_3$/l.[2] However, there remain persistent hot spot areas across Europe at the regional level with nitrate levels well above these averages, and the need to increase the uptake of measures in these areas is paramount to the continued sustainability of drinking water extraction.[3] Several countries in Europe report aquifers having concentrations of pesticides that exceed the Environmental Quality Standard (EQS). EQS is water quality standards for 33 priority substances and 8 other pollutants defined in Directive on Environmental Quality Standards (2008/105) amended by a directive (COM(2011)876) for another 15 priority substances. Despite the increased integration of policies to deliver clean and safe drinking water over the last 30 years there is clearly a need to increase the engagement between interdependent actors and stakeholders.[4] The reduction of the diffuse pollution of drinking water sources by pesticides and fertilizers used by the agricultural sector remains the biggest challenge and requires a move toward more "horizontal" water governance between the various actors and stakeholders: water companies, farmers, nature conservation NGOs, plant protection product (PPP) producers, fertilizer producers, food and retail businesses, consumer organizations, environment agencies, and ministries.[5] From a regulatory perspective, there is a clear need to obtain more engagement of actors in the implementation of Directive 2009/128/EC which aims to achieve a sustainable use of pesticides in the EU. The main actions that are taken in national action plans relate to training of users, advisors and distributors of pesticides, inspection of pesticide application equipment, the prohibition of aerial spraying, limitation of pesticide use in sensitive areas, and information and awareness raising about pesticide risks. Relatively little attention is paid to the socioeconomic aspects related to sustainable use and the financing of alternative farming systems toward reduced risks.

The objective of this chapter is to describe methodologies and technologies that contribute to the effective uptake and realization of innovative farming systems and mitigation measures delivering good water quality with a specific focus on agrochemicals. Therefore, we describe here an integrative multiactor participatory framework, based on four key aspects: (1) innovations in water governance, (2) participatory monitoring, (3) farming systems and management practices, and (4) software tools that enable actors to effectively implement management practices and measures for the protection of water sources.

## 2. GENERAL METHODOLOGY

We propose to conceptualize water resource systems as complex socioecological systems.[6,7] The geographical boundary of the water resource system is generally the catchment boundary as shown in Fig. 1. Water

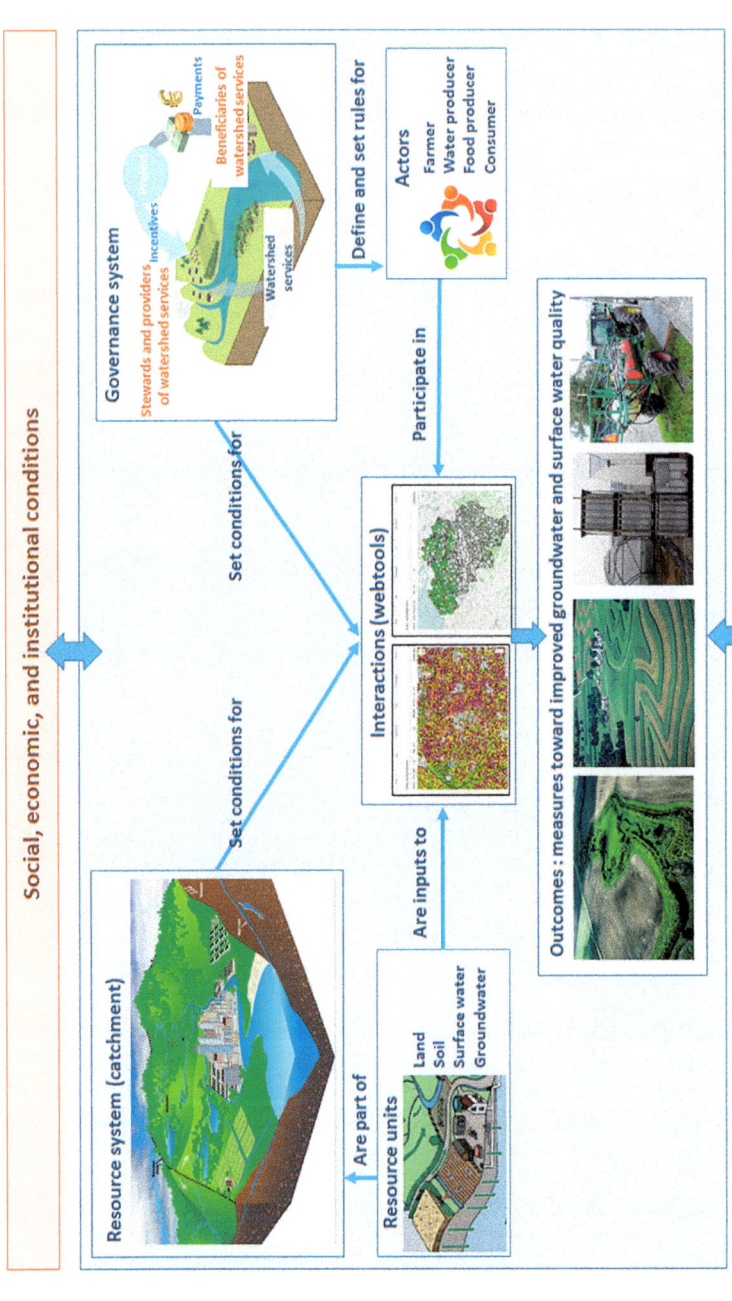

**Fig. 1** Conceptualization of water resource systems as complex socioecological systems.

systems can be conceptualized as socioecological systems in which different users manage the resource system through their actions and interactions. There is a clear need for a multiactor approach in the design of effective solutions. Indeed, given the size and nature of water-related problems, tackling them requires a coordinated effort among innovators, policy makers, and stakeholders.[8] Actors together with experts from different disciplines, such as chemistry, hydrology, hydrogeology, engineering, information technology, sociology, agronomy, farming systems design, and economics, must jointly develop innovative instruments that facilitate the uptake and impact. The central question is to find the appropriate and effective institutional arrangements that can focus how actors behave and cooperate, in order to achieve the preservation and improvement of groundwater and surface water quality.

We propose to organize the multiactor approach in so-called action labs headed by an action lab leader who stimulates a good cooperation between stakeholders, a greater involvement of farmers and citizens in the monitoring of water quality, and the adoption and long-term durability of efficient on-farm land-use strategies. The action lab leader preferably has established a long-term and trustful relationship to the key actors. Key actors in the frame of sustainable use of agrochemicals are the farmers.

## 3. THE MULTIACTOR APPROACH

### 3.1 What Is a Multiactor Approach?

A multiactor approach creates conditions for interaction between different actors involved and combines their knowledge, perspectives, resources, and experiences to identify and discuss solutions and new ideas.[9,10] Researchers, farmers, entrepreneurs, governmental organizations, etc. all have different forms of knowledge (practical, scientific, policy based, etc.). In the multiactor approach, we give a voice to all actors so that we can incorporate the different forms of knowledge in the discussions. We create a setting in which people with different interests can discuss and cooperate. In that way, the actors can negotiate about goals, decision making, and activities. When we talk about a multiactor approach, we also speak of a cross-pollination of knowledge, cocreation of results, or coownership of results.

### 3.2 Why a Multiactor Approach?

More thoughtful, innovative, and broadly accepted decisions can be developed by applying a multiactor approach.[11] A broader view on who will benefit or who will be harmed by a decision will be obtained by including the different perspectives and interests in the process. Such an approach of bottom-up and inclusive decision making is the key to effective (water) policies.[9] By increasing the awareness of specific problems and acceptance of policy actions among the different actors, a reduction of the occurrence of unexpected resistance can be expected.[12] Solutions will be better adapted to context specific circumstances when they are developed and discussed with the local stakeholders. The approach also aims for a good cooperation between the stakeholders.

More and more projects include a multiactor approach to reach innovative and accepted ideas. The need for a multiactor approach results from the observation that many experts are developing technical solutions for specific problems; however, they are not always applied. Up until now, the management of many environmental resources is mechanistic and technocratic and often neglects complexity and the human dimension. Another observation is that many societal problems cannot be solved by just one actor. Neither the state nor the private sector can act alone to solve socioecological issues such as water quality related to agriculture.

## 3.3 Planning of the Multiactor Approach in Action Labs

By setting up a multiactor approach in the framework of water quality and agriculture, different voices will be heard and there will be a continuous alternation between the theoretical knowledge of the researchers and experts and the more practical and policy-related knowledge of other actors involved (farmers, farmers advisory, conservationists, water producers, agricultural ministries, and environment agencies). This multiactor approach should be part of every aspect that is needed to engage actors to improve water quality: water governance, participatory monitoring, best management practices (BMPs), and collaborative tools (Fig. 2). In what follows, we describe some practical steps for the setup of the multiactor approach.

### 3.3.1 Step 1: Identifying the Actors

We involve all actors that have an influence on or that are influenced by the water system. In a first step, we gain more insight on who these actors are and who they represent. This can be an individual, a client, a decision maker, a Non-Governmental Organization (NGO), a business or corporation, a politician, or a community. We will identify the different actors together with their interests, resources, and expectations related to the water quality issue. In what follows, we give a method to make an overall overview of the actors. This gives the basis for the further process.

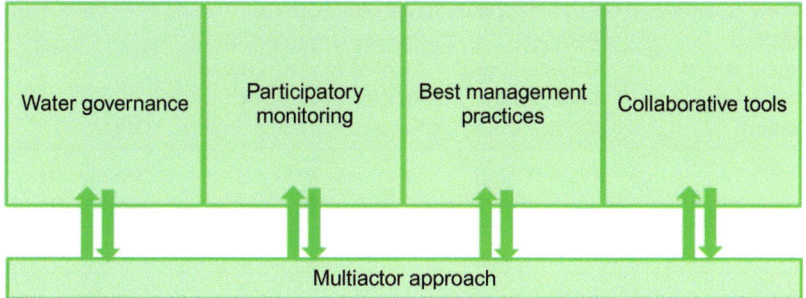

**Fig. 2** Multiactor approach in each of the four key aspects related to water quality and agriculture.

### 3.3.1.1 Data Collection

First of all, we have to collect data about the actors that have an influence on or that are influenced by the water system. Based on desk research, previous experiences, or exploratory conversations with some key actors in the region, an overview of the actors and organizations active in the catchment can be made. We are looking for more information on the different actors in the region and get to know their stances related to the water quality. Following example questions help to find as much relevant actors and organizations as possible:

- Government (agricultural ministries, environmental agencies, etc.)
  o What sectors have an influence on the policy on water issues?
  o What public authorities represent these different sectors? (on local, regional, national, or international level)
- Companies (farmers, farmers advisory, water producers, fertilizer, and PPP producers)
  o What companies are present in the case study area?
  o Are the companies represented by an organization?
  o Are there farmer cooperatives present in the case study area?
- Nonprofit and social profit organizations (conservationists, civil society, NGOs, etc.)
  o Are there environmental action groups active in the case study area?
  o Are there citizen organizations active in the case study area?

### 3.3.1.2 Data Interpretation

In the interpretation step, we map the stakeholders in a structured way. As stakeholders will be very diverse, it can be helpful to define groups of stakeholders. A possible categorization of the stakeholders is given in Table 1.

**Table 1** Example of Categorization of Stakeholder Groups

| People/ Organizations Involved in Farming Systems | People/ Organizations Involved in Water Issues | People/Organizations Involved in the Linkage Between the Agricultural Sector and Water Policy | Other |
|---|---|---|---|
| Farmers | National authorities | Local authorities | Researchers |
| Farmers' unions | Water utilities | Advisors | Conservationists |
| Extension services | Policy makers | NGOs | Citizen organizations |
| Advisors | | | Entrepreneurs |

For every actor, we want to get more insight in their position, their interest in the project, and their resources. Based on interviews with the key actors in the area, the position of the actor to the water system or water body, to the project (i.e., improvement of water quality), and the resources related to water quality issues can be derived. Aspects related to the position of the actor to the water system are: regulatory issues, water use, advisory roles, responsibilities. Questions, such as: "Why would stakeholders be willing to cooperate? What's in it for them? Will they support or resist the project?," allow to derive the position of the actor to the project. Resources to implement the project depend on the financial benefits and position of the actor, his knowledge, the production environment, the legitimacy of the project, and the power to influence the project. Points of attention in the actor identification are: a great diversity in perspectives, priorities, etc. within a single group (for example, farmers), to avoid preconceived ideas and assumptions about the different actors, to not communicate information to other people if it is not checked by the actors themselves. The interpretation is not always neutral and can damage confidence.

The actor analysis is a snapshot of a certain moment in time. It is possible that new actors enter the region or that positions and interests of some actors change during the course of the project.

### 3.3.2 Step 2: Designing the Process

This section presents a summary of guidelines on how to organize the process of the different interactive moments ranging from information sessions about the project to one-on-one interviews on individual

**Table 2** Designing the Process: Overview of All Interactive Moments for the Key Aspects Related to Water Quality

| Key Aspect | Timing Goal | Who | Method | Input | Output |
|---|---|---|---|---|---|
| Governance Participatory monitoring Best management practices Collaborative tools | What are the goals? | Which actors need to be involved? | Individual, focus groups, brainstorm, etc.? | Which input from previous steps is required? Which predefined input is required | What do we want to achieve? Information for which aspect? |

perspectives, to stakeholder discussions for problem identification, goal setting, responsibility sharing, design of data collection, and deciding on collective actions. Each step of this process needs to be carefully discussed and deliberated. Every next step in the process usually depends on the output of previous steps, therefore the design is variable and the process can be very complex.

### 3.3.2.1 Overview of the Activities

It is important to have one manager to keep an overview over the whole process (in most cases the action lab leader). Based on the required information for the key aspects related to water quality (governance, participatory monitoring, BMPs, and collaborative tools), the action lab leader makes an overview of all stakeholder involvement moments, their associated actors, the goal, and the required input and output (Table 2). To avoid taking too much time and engagement from the stakeholders, this overview can help to combine different stakeholder moments into the same interactive session.

### 3.3.2.2 The Use of Different Methodologies

The action lab leader can choose between different techniques to collect the data: in-depth interviews, focus groups, questionnaires, information moments, or flyers. The type and level of actor involvement needs to be flexible and adaptable to changing circumstances. The methodologies used in the process can be qualitative or quantitative. To explore the perspectives on a particular situation of an actor, the qualitative research technique "in-depth interview" can be used. By seeing the actors separately, one gets the chance to collect information about different themes, but also about their worries, concerns, and ideas. A focus group is a method to gather information from a group of people. The researcher selects the topics, while the focus group participants provide the data. Focus groups produce large amounts of concentrated data in a short period of time. Questionnaires or surveys can be drawn up to gather information in a more quantitative way.

Implementing these methodologies requires specific expertise. It is important that action lab leaders who do not have experience in multiactor processes and the related methodologies collaborate with a partner in the action lab who does have the required expertise.

More information on qualitative interviews can be found in Baxter and Eyles.[13] More information on the planning and implementation of focus groups can be found in the "Focus Group Guidebook" by Morgan.[14] More information on qualitative research and methods in general can be found in Creswell[15] and Patton.[16]

## 3.4 Guidance for the Implementation of a Multiactor Approach

The action lab leader coordinates and implements the process. He should be unbiased and not influenced by other actors. He needs enough flexibility to adapt the process to the local context and circumstances. The following tips give some guidance for a successful implementation of the multiactor approach. They are based on the five components to organize the social interface of rural policy making that have been developed by Rogge et al.,[10] combined with insights from other authors.[9,17,18]

### 3.4.1 A Transparent and Fair Process

Give adequate and accessible information on the project and the course of the process. Inform the actors why and how they are consulted at a specific moment in time. Actors need reliable information about their legitimacy in decision making: what will be done with the input from the actors? To what degree can the actors steer the process and decisions? What will happen with their data? Procedures should be applied consistently and fairly across actors over time.

### 3.4.2 Visualization of the Process for Better Understanding

The leader needs a clear view on how the process is organized. A useful tool to manage the process and to inform the participants is a visualization of the process design in a scheme. The different steps of data gathering, data processing, and stakeholder discussion can be visualized. In such a scheme, it can become clear when different groups of actors are involved and for what steps the obtained information is used. An example of the visualization of a multiactor process is given in Fig. 3.

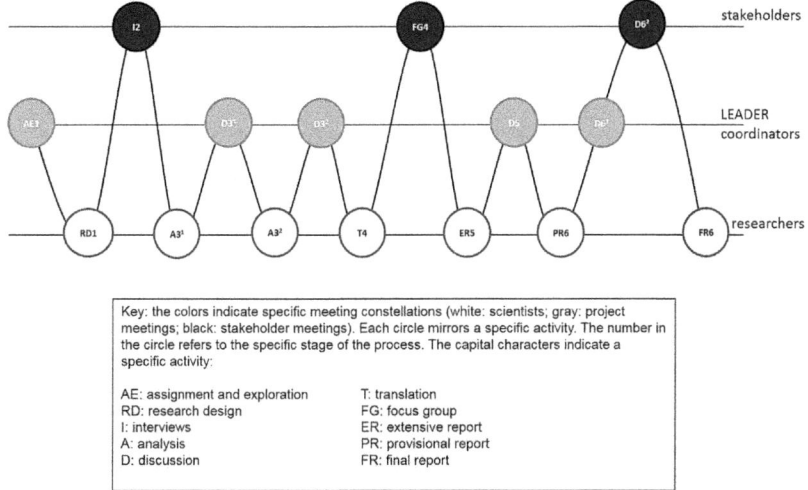

Key: the colors indicate specific meeting constellations (white: scientists; gray: project meetings; black: stakeholder meetings). Each circle mirrors a specific activity. The number in the circle refers to the specific stage of the process. The capital characters indicate a specific activity:

| | |
|---|---|
| AE: assignment and exploration | T: translation |
| RD: research design | FG: focus group |
| I: interviews | ER: extensive report |
| A: analysis | PR: provisional report |
| D: discussion | FR: final report |

**Fig. 3** Visualization of the process design to develop a Local Development Strategy (LDS) in LEADER areas in Flanders.[18]

### 3.4.3 An Equal Involvement of All Actors

Every actor affected by a decision should have the possibility to join the process and to make their perspectives heard. In a first meeting, all relevant actors have to be invited so that they have the opportunity to participate in the whole process. As an action lab leader, one has to be aware of power imbalances and overrepresentation of some actors. There has to be a balance in the consultation of different actor groups. Action lab leaders have to assure an equal participation of the stakeholders. If there are imbalances, the action lab leader can intervene in the multiactor process by approaching actors separately to encourage disinterested people or possible blockers. The multiactor approach is an open approach, actors can (temporarily) join or leave.

### 3.4.4 A Neutral Start for the Process by Sharing Common Objectives and a Common Language

Define common goals and a common shared language at the start of the process. Common objectives and desired outcomes actively engage the participants. A shared language provides a neutral starting point that the diverse actors can all agree on. A balance between technical and nontechnical information is required. All actors share the power to define and agree on the common understanding of the issues at stake and the defined objectives. The action lab leader should create an environment of trust and security so that every actor can speak openly and truthfully.

### 3.4.5 Social and Emotional Dynamics to Encourage Overall Group Functioning

It can be interesting to incorporate activities that stimulate the group's overall functioning next to the activities that serve the needs of the project tasks. This could create more solidarity within the group and benefit the overall functioning of the group. Be aware that participatory processes are resource consuming.

In the following paragraphs the key aspects of the multiactor approach are further highlighted: water governance, participatory monitoring, best management, and collaborative tools.

# 4. WATER GOVERNANCE

## 4.1 General

The issue of water governance systems has received considerable attention by such international organization as the Organization for Economic Cooperation and Development (OECD) and the GLOBAL Water Partnership (GWP). The GWP defines water governance as the range of political, social, economic, and administrative systems in place to develop and manage water resources, and the delivery of water services, at different levels of society.[9] This attention has led to a large number of guidelines, principles, and frameworks for water governance systems. Yet, there are still gaps in understanding why certain governance systems work effectively and efficiently in practice and others do not. Here, we present a framework for governance related to water quality and agriculture. The framework is based on academic[12,19,20] and nonacademic literature.[8,9] The framework identifies the attributes that are imperative for good governance systems. Pahl-Wostl[12] identifies formal and informal institutions, actor networks, multilevel interaction, and governance modes as key attributes for successful governance systems. This stakeholder engagement is operationalized in the multiactor groups with influenced and influential actors who will have continued participation in the action labs. Other attributes include the degree and forms of collaboration, ownership conditions, and the type of incentive structures. In governance systems, collaboration between stakeholders is often on a voluntary basis, which defines the need for appropriate incentive structures (transition pathways), from paying for depolluting to payment for ecosystem services or other benefits. Knowledge exchange on BMPs, the setup of (participatory) monitoring systems and the collaborative data platform will support or be part of this governance system.

The governance system defines and sets the rules for the way actors collectively behave. In the frame of sustainable use of agrochemicals, the resource system is the catchment, which is formed by the resource units land, soil, surface water, and groundwater. The catchment area is managed by different users through their actions and interactions. Water governance refers to the range of political, social, economic, and administrative systems (including the institutional setup and the organization of the drinking water supply system) that are in place to develop and manage water resources, and the delivery of water services, at different levels of society.[9] The governance of the action labs can be understood as the rules and arrangements between all actors involved to deliver drinking water as well as prevent its pollution.

In many European countries, the governance of water systems is characterized by an expert culture based on technological solutions, a lack of integration of different information sources and rigid regulations.[12] However, several people, organizations, and institutions are involved in water management. This complexity is little or not at all embedded in the ongoing management and governance of the water system. As a result, implementation of BMPs remains poor in practice, even though the benefits are known. These considerations must be kept in mind in order to cope with the challenge to provide good water quality. Well-functioning governance systems have to be developed that incorporate all relevant actors and that can deal with the complexity and uncertainty of the water systems. Water governance research will focus on both administrative systems, with a focus on formal institutions (laws and policies) and informal institutions (power relations and practices) which different actors use to achieve their objectives of the preservation and enhancement of the surface- and groundwater quality. The governance research aims to determine characteristics, constraints, and possibilities to develop governance models that support water management systems at the local level, which enhance the quality of ground- and surface water.

## 4.2 Water Governance Framework for Water Quality and Agriculture

This section describes the governance framework for analyzing and improving water governance systems with the aim to effectively implement measures to improve water quality related to use of agrochemicals.

The framework consists of three steps (Fig. 4):

**(1)** The water system and the specific issue of (drinking) water quality (central part in Fig. 4). The focus here is on the influence of pollutants

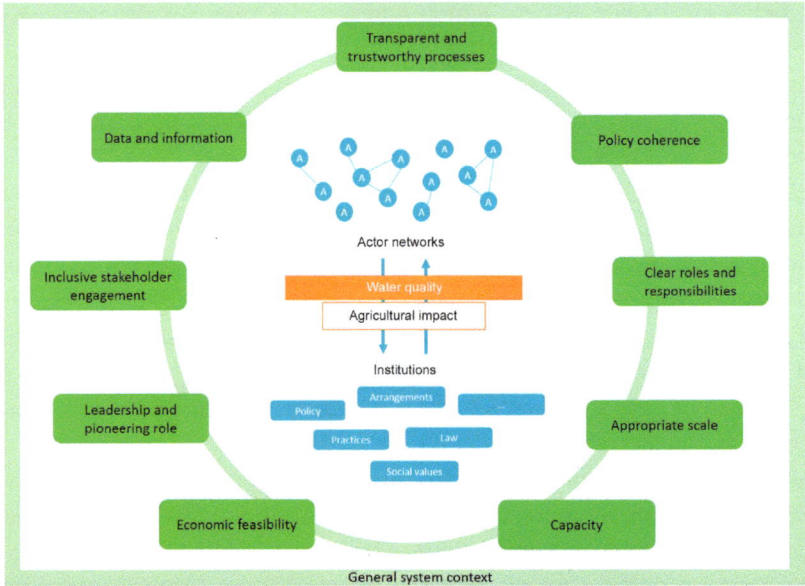

**Fig. 4** Framework for water governance: water system (*orange*), governance system (*blue*), building blocks for a constructive environment for development of a good and well-functioning governance system (*green*).

produced by human actions and interactions, and more specifically from point and diffuse sources in agriculture.

**(2)** The components of the governance systems (above and below central part in Fig. 4). The governance system exists of the interactions between actor networks and the institutions they formulate. The proposed arrangements/solutions trying to cope with the impact by the agricultural sector on the water quality are described, in order to analyze whether these solutions can fully cover this impact.

**(3)** Building blocks to obtain a "constructive environment for development of a good and well-functioning governance system" (surrounding circle with blocks in Fig. 4). The presence of these elements can increase the degree of implementation of the proposed solutions. These building blocks should guide the user of the framework to diagnose failures in their governance structure and to give an indication of the elements where improvements could be made.

The framework is expected to serve as a tool to analyze and improve water governance systems in action labs. There are some challenges for the development of such a framework. On the one hand we know that there are no one-size-fits-all answers for governance challenges all over Europe.[5] Therefore, the framework has to be generic enough so that it can be applied across different biophysical, social, economic, and political conditions. On the other hand, the framework has to be specific so that it can take into account the complexity, specificities, and concerns of the water quality issue in each action lab. The insights of similar problems and contexts should be transferable. Another challenge is to make the framework very practical so that it is easily applicable in the action research, but also in future practices.

## 4.3 Overview of the Governance Framework in Three Steps

### 4.3.1 Step 1: The Water System and Related Problems: Focus on Agricultural Impact on Drinking Water Quality

River basins are inherently complex systems with many natural and socio-economic dynamics across spatial and temporal scales. Hence, in order to understand the functioning of the river basin, it is required to study both the dynamics of the natural system and the socioeconomic dynamics.[21] The focus here is on the issue of the (drinking) water quality, which is one element of the water resource system. It should always be kept in mind that there are many more issues related to the water resource system (e.g., water quantity).

The issue of (drinking) water quality can be made more specific. The type of pollutants and the type of water source under study will differ in each action lab or case study. It is important to specify the problem in detail in each specific action lab, but keep in mind that there are many pollutants influencing the drinking water quality in both surface and groundwater. The level of pollutants is influenced both by natural and human processes. In this water governance context, we focus on the influence of pollutants produced by human actions and interactions, and more specifically from point and diffuse sources in agriculture (Fig. 5).

The territorial characteristics of the river basin and the agricultural sector will differ across the action labs. It can be useful to gain more information on these characteristics to provide insight in the type of river basin for which a water governance system has to be set up.

Questions related to the water system and the related challenges are:

- What are the purposes of the river basin in your action lab (drinking water production, wastewater treatment, hydropower, irrigation, environmental management, tourism)?

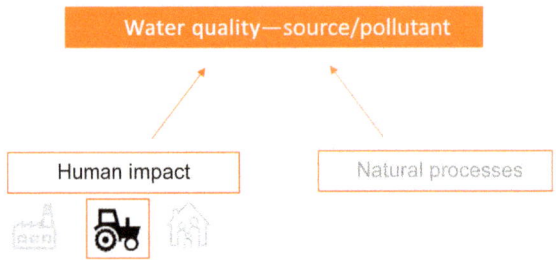

**Fig. 5** Step 1 in the governance framework: defining the water system and related problems. The focus is on the human impact of the agricultural sector on the system of the (drinking) water quality (both surface water and groundwater).

- Which source (ground- or surface water) is under study in your action lab?
- What are the water quality objectives in your action lab defined by law?
- Which type of pollutants are present in your action lab?
- Which type of pollutants will be under study in your action lab?
- What is the impact of natural processes on the level of pollutants?
- What is the impact of the human actions on the level of pollutants?
- What is the typical farm structure and organization in your action lab?
- What are the main production outputs?
- What is the impact of the agricultural sector on the level of pollutants?
- What are the causes of the agricultural impact (pollution by point or diffuse sources, etc.)?
- What are the challenges to cope with the agricultural impact on the water quality?

### 4.3.2 Step 2: The Governance System of the Water Quality: Interactions in the Actor Network and Institutions

Fig. 6 gives the schematic view on the components of the water governance system: actor networks and formal and informal institutions.

#### 4.3.2.1 Actor Networks

Actor networks emphasize the roles and interactions of the different state and nonstate actors who have an impact on water quality. The impact of different actors on the water quality is complex in different ways. First of all, the water body is managed and influenced by many different actors through their actions and interactions. Both public, private, and civil society actors influence or are being influenced by the water quality. Second, the water quality issue is strongly linked to different sectors such as environment,

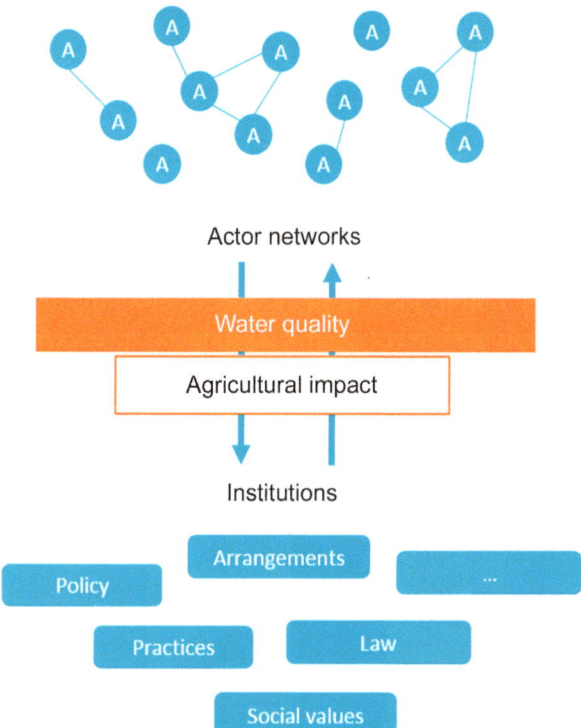

**Fig. 6** Step 2 in the governance framework: schematic view on the components of the governance system.

health, agriculture, spatial planning, energy, and public works/infrastructure. These many policy areas have their own organizational culture and different degrees of sensitivity to lobbies. Third, the water quality management is both a global and local concern resulting in interdependencies across different levels of government (local, regional, national, and supranational). In the last few decades, this trend has been exacerbated by the increasing involvement of both local (for whom water is a local concern) and supranational actors (for whom water is a global concern).[5]

All these actors at different administrative and territorial levels share the responsibility to deal with the water quality in a suitable way. In order to cope with these challenges, solutions should be designed and implemented on multiple levels and involve multiple actors and multiple sectors.[10] Further, the kind of interactions and interdependence among the actors is an important factor.

Questions related to the actors and their specific roles are:

Describe for each of the actors their specific role in regulating or influencing the drinking water quality. Actors are all people or organizations who have an influence on or are influenced by the water quality in the action lab. At least the following actors should be mentioned: local government (municipalities), regional government, national government, supranational government, water producers/drinking water suppliers, wastewater treatment facilities, farmers, environment/river basin agency, farmers advisory and farmers' union, local businesses, relevant companies operating as food traders, corporate buyers, chemical producers, processors, and retailers that can have influence on farmers, civil society organizations, research organizations, consumer organizations, consumers, etc. If one or more of these actors is not involved in the current water governance system, please mention this.

- What is their role/responsibility related to the water quality (priority setting and strategic planning, allocation of uses, economic and environmental regulation, information and monitoring, infrastructure operation and investment, advising, implementation good practices, etc.)?
- How is the current water quality perceived by the actor? What causes this perception?
- How could the actor benefit from a good water quality?

Describe the relationships between the actors mentioned earlier.

- Which actors interact with each other about the water quality?
- How do they interact and with what frequency?
- What are the ambitions for the partnership/interaction?
- Is the collaboration formalized via a contractual agreement of some sort?
- What is the relation between public authorities and private (local) actors?
- What is the relation between actors from different sectors?
- What is the relation between actors from different government levels?

### 4.3.2.2 Formal and Informal Institutions

Actors design and implement institutions or (a set of) rules that (are supposed to) govern their behavior.[22] A distinction is made between formal and informal institutions, which refers to the nature of processes of development, codification, communication, and enforcement.[12] Formal institutions refer to rules that are defined in regulatory frameworks and can be enforced by legal procedures, like, for example, laws or official policies. Informal institutions refer to socially shared rules such as cultural norms, values, and belief systems. These rules are often not written down and are enforced "outside of legally sanctioned channels".[12] Examples are power relations and practices. Both formal and informal institutions play an important role in water resources governance: their potential to set rules and demarcate responsibilities between actors, their coordination of mechanisms to minimize jurisdictional overlaps or deficiencies, the bridging of the gap between political and natural boundaries, matching responsibilities, and serve as authorities and facilitators of action.[23,24]

Questions related to institutional arrangements are:

Describe the formal policies/legislation that aim to improve water quality and prevent pollution.

- What are the most important policies/legislation (on international, national, regional, or local level) in your action lab to improve the water quality? What are the goals? Are there any goals going beyond the goals setup in the EU legislation?
- Which legal procedures related to water quality (for instance EU and national legislation, governmental directives, river basin management plans, official partnership agreements and rules, etc.) exist in your action lab or river basin?

Describe the economic incentives that aim to improve water quality and prevent pollution.

- Are there specific payments (e.g., pollution charges, payment for environmental services, solidarity mechanisms, investment plans and programs, rural development fund, agrienvironmental schemes, etc.) that aim to improve water quality and prevent pollution?

Describe the social incentives that aim to improve water quality and prevent pollution.

- Is there specific support provided (farm advice services, sensitizing campaigns, education programs, etc.) that aim to improve water quality and prevent pollution?
- How is social learning to facilitate dialogue and consensus-building promoted (e.g., networking platforms, social media, user-friendly interface (digital maps, big data, smart data, and open data), etc.)?

Describe the innovative actor arrangements that aim to improve water quality and prevent pollution.

- Which innovative water arrangements (e.g., intermunicipal collaboration, urban–rural partnerships, performance-based contracts, etc.) exist in your action lab?
- How are experimentation and pilot-testing on water governance facilitated or encouraged in your action lab?

Describe the good practices that aim to improve water quality and prevent pollution.

- Which BMPs are considered relevant for good water quality? How will the implementation of these systems increase the water quality?
- To what extent are they currently being implemented? Why (not)?

Describe the communication programs to improve water quality and prevent pollution. How is the communication about (issues on) the water quality organized?

- Are there regular fairs, events, or community places where people meet and exchange information about the water quality?
- What communication tools and infrastructures are used by farmers in their learning practices? (Internet, e-mails, electronic advice provision, e-subscription lists, newsletters, online training activities, etc.).

### 4.3.3 Step 3—Building Blocks for a Constructive Environment for the Development of a Well-Functioning Water Governance System at Local Level

A constructive environment can help in order to set up a well-functioning water governance system, therefore the presence of a number of building blocks is required. These can be considered as the skills of the environment (leadership, information, transparency, etc.), required for the succeeding of the formal and informal institutions.

Fig. 4 gives a visualization of the framework for good and well-functioning water governance (circle with blocks).

#### 4.3.3.1 Data and Information

Data and information about the state of the water quality, related policies, and processes should be available in order to improve the understanding of the water quality issue and the effectiveness of measures taken to improve water quality. Therefore, monitoring campaigns should be set up. Clear agreements about sharing data should ensure availability of results to a wide audience. It is important that this information is consistent, up to date, and shared timely by all actors. Terminology related to water quality and methodology of measurements should be clear and unambiguous. In practice, subnational governments and other local actors will tend to have more information about local needs and preferences, and also about the implementation and costs of local policies, while central governments play a role in managing the information so as to support a broader vision of public policy objectives.[5] In some cases, companies do not supply data when requested, in order to avoid the subsequent additional restrictions that would occur on the basis of the information in the data.[5]

A gap occurs when there is an asymmetry of information across different actors or a lack of capacity, resources, and expertise to collect, analyze, and interpret water data.

Related questions are:

Describe the measures taken in the governance system in your action lab in order to ensure data collection.

- What type of (water quality related) monitoring currently exists in your action lab? Who takes the responsibility for doing this?
- How does the monitoring contribute to the objective of improving the water quality?
- Is (a part of) this monitoring participatory? In what way? To what extent? Are citizens involved?
- What actions are taken to avoid overlaps, synergies, and unnecessary data overload?

Describe the measures taken in the governance system in your action lab in order to ensure data sharing.

- What monitored data are available to the public? How are information and knowledge on the water quality made available to the public? Who takes the responsibility for doing this?
- How are data and experiences shared among different organizations and agencies producing water quality data? Who takes the responsibility in doing this?
- What strategies are used to share the results as new information becomes available in a timely and transparent manner? Which channels are used to inform about the existence of new data?

### 4.3.3.2 Transparent and Trustworthy Process

Transparency in design and implementing processes could reduce the chances of corruption. Decisions related to the water quality should be transparent, so that relevant actors can follow and join the process at any time. The actors should be informed why and how they are consulted at a specific moment in time and need reliable information about their legitimacy in decision making: what will be done with the input from the actors? To what degree can the actors steer the process and decisions? What will happen with their data?

Trust enables people to go beyond their own personal, institutional, and jurisdictional frames of reference and perspectives toward understanding other peoples' interests, needs, values, and constraints.[25–27] Trust grows over time as parties work together, get to know each other, and prove to each other that they are reasonable, predictable, and dependable.[11,28] Trust generates mutual understanding (the ability to understand and respect others' positions and interests even when one might not agree), which in turn generates legitimacy and finally commitment to a shared path.[11]

A gap occurs when there is a lack of legitimacy in decision making and a lack of trust.

Related questions are:

Describe the measures taken in the governance system in your action lab in order to ensure trust.

- What is the level of trust among the involved actors?
- Which incentives are present in your action lab to improve trust among actors?
- Which incentives are present in your action lab to mitigate conflicts among different actors?

Describe the measures taken in the governance system in your action lab in order to ensure transparency.

- What is the level of transparency for decision making on water quality?
- Which norms, codes of conduct, or charters on transparency exist in the design or implementation process of the formal and informal institutions?

### 4.3.3.3 Policy Coherence

Policy related to multiple levels and multiple sector should be coherent and can be ensured by coordinations mechanism. The "rules of the game" need to be clearly spelled out, as should be the consequences for violating the rules, and arbitration enforcing mechanisms should be built in to ensure that satisfactory solutions can still be reached when seemingly irreconcilable conflicts arise among the stakeholders.[9] Conflicting interests could occur among actors, but also between rural and urban areas, and upstream and downstream areas.[5]

A policy gap occurs when there is a sectoral fragmentation of the tasks across different actors.

Related questions are:

Describe the measures taken in the governance system in your action lab in order to handle with different interests and objectives.

- How coherent are the existing solutions to cope with water quality and prevent pollution? Do they contradict each other?

- Which conflicts of interest in water quality among actors are present in your action lab?
- Which measures are taken to cope with different interests and objectives?
- To what extent/how is coherence stimulated?

### 4.3.3.4 Inclusive Stakeholders Engagement

All actors affected should have the possibility to join the process and to make their perspectives heard. The multiple perspectives and different interests of the actors give a broader view of who will benefit or be harmed by actions taken.[11] Equity between and among the various actors needs to be ensured throughout the process of development and implementation of water policy. Above all, water governance has to be strongly based upon the ethical principles of the society in which it functions and based on the rule of law.[9] Through stakeholders' engagement, people with differing perceptions, relational, and identity goals work across their respective institutional, sectoral, or jurisdictional boundaries to solve problems, resolve conflicts, or create value.[11] Stakeholders' engagement should be ensured throughout the whole process, from conception to implementation.[9] It is an open process where actors can join or leave the process at any time.

A failure in inclusiveness of all actors can be defined as the existence of power imbalances in the process and the presence of over- or underrepresentation of some actors.

Related questions are:

Describe the measures taken in the governance system in your action lab in order to ensure inclusive stakeholder's engagement.

- To what extent are all relevant actors engaged in the decision-making process?
- To what degree and in what ways are different policy areas (esp. agricultural policy, environmental policy, etc.) integrated within your action lab?
- To what degree are actors from one level (e.g., basin level, the national level) involved in decision processes at another level (e.g., the European level)
- Which institutional frameworks, organizational structures, and responsible authorities are conducive to stakeholder engagement?
- Who are underrepresented actors? How are underrepresented actors motivated to join the process?
- Who are overrepresented actors? How are power imbalances and risks of unilateral information from overrepresented actors mitigated?

### 4.3.3.5 Clear Roles and Responsibilities

Coping with the water quality raises the question of "who does what" and "why." Roles and responsibilities need to be clear in the processes of policy making, policy implementation, operational management, and regulation.[8] In that way, each actor or organization knows what is expected of them and can take responsibility for what it does.

There is a gap when there is a lack of clarity on roles and responsibilities, a hyperfragmentation of roles and responsibilities, or overlapping roles across different actors. Diverging or contradictory objectives between actors will obstruct long-term planning for integrated water management.

Related questions are:

Describe the measures taken in the governance system in your action lab in order to ensure clear roles and responsibilities.

- What are overlapping or unclear roles or responsibilities in the governance system (priority setting and strategic planning, allocation of uses, economic and environmental regulation, information and monitoring, infrastructure operation and investment, evaluation, implementation, etc.)?
- How are the responsibilities of the different actors defined (by law, by arrangements, etc.)?

### 4.3.3.6 Appropriate Scale

Water quality should be managed at the appropriate scale to reflect local conditions, and foster coordination between different scales and sectors. In the water sector, the administrative boundaries of municipalities, regions, and states rarely correspond to hydrological boundaries. This often results in management failures. For example, it is not logical to enforce different water quality regulations for different sections of one river due to administrative or political fragmentation.

An administrative gap exists when there is a geographical mismatch between hydrological and administrative boundaries.

Related questions are:

Describe the measures taken in the governance system in your action lab in order to ensure operation at an appropriate scale.

- At what scale are plans made and measures taken to improve the water quality?
- Which methods are applied to ensure that basin management plans are consistent with national policies and local conditions?

### 4.3.3.7 Leadership and Pioneering Role

Leadership refers to the presence of an identified leader who is in a position to initiate and help secure resources. The leader should possess a commitment to collaborative problem solving, a willingness not to advocate for a particular solution, and exhibit impartiality with respect to the preferences of participants.[29,30]

Related questions are:

Describe the measures taken in the governance system in your action lab in order to ensure leadership.

- Which roles of leadership are present in your action lab (sponsor, facilitator/mediator, science translator, public advocate, etc.)?
- Who is responsible for the leadership roles in your action lab?
- Which measures are taken to stimulate leadership?

### 4.3.3.8 Capacity

The capacity of the actors is defined as the whole range of the competences required to carry out one's duties related to the water quality.[5] These relate to among others organizational, technical, procedural, networking, or infrastructure capacity. A common problem related to capacity is the limited capacity at the local level.[5] In the past, many countries have allocated complex and resource-intensive competences from higher levels of government to lower levels of government.[5] The subnational actors do not always have enough financial and organizational capacity required to meet the needs to design and implement suitable water policies. This may lead to the deterioration and potential failure of services and infrastructure, which in turn threatens the quality of water resources. Established subnational governments with well-developed institutions may need little capacity building when faced with new responsibilities. Cases, where local governments or actors historically had a limited role in water quality issues, will encounter greater difficulties.

A capacity gap is defined as the mismatch between the level of technical, financial, or institutional capacity and the nature of the problems and needs.

Related questions are:

Describe the measures taken in the governance system in your action lab in order to ensure the capacity of the actors.

- Which technical, logical support, and administrative and organizational assistance is present in your action lab in order to strengthen the capacity of the actors?

- Which trainings are organized to learn and improve skills/competences needed in the design and implementation of water policy and arrangements?

### 4.3.3.9 Economic Feasibility

The maintenance of a good water quality requires a sufficient long-term investment. Adequate and stable budget should ensure the implementation of water policies and arrangements among different actors. There could exist different arrangements that help actors raise the necessary revenues to meet their mandates, building on, for example, principles such as the "polluter pays" and "user-pays," as well as payment for environmental (ecosystem) services. Private partners, investment banks, and innovative arrangements at local level could complement public action in water financing.

A funding gap refers to insufficient or unstable revenues for implementation or to a situation where there is a difference between actors' revenues and the expenditures actors require to meet their responsibilities. Previous research of OECD stated that subnational authorities often depend on higher levels of government for funding water policies, while these higher levels of government depend on the former to deliver them information and meet both national and subnational policy priorities.[5]

Related questions are:

Describe the measures taken in the governance system in your action lab in order to ensure financial feasibility.

- Which mechanisms (e.g., social contracts, scorecards, audits) exist that foster efficient and transparent allocation of water-related public funds?

### 4.3.3.10 General System Context

The general system context describes the context of governing style, traditions, and values which characterizes a certain territory. These more general cultural and social systems will be different in each country and can create opportunities or constraints to the way actors behave in relation to the water quality.[11] Elements that are important to get insight in the general system context are the general appreciation or cultural traditions related to water resources; the general resource conditions in need of improving, increasing, or limiting; prior failures to address the issues through conventional channels and authorities; historic levels of conflict among recognized interests and the resulting levels of trust and impact on working relationship; policy and legal frameworks, including administrative, regulatory, or judicial; political dynamics and power relations within communities and among/across levels of government; degree of connectedness within and across existing networks and socioeconomic and cultural health and diversity.[11]

Related questions are:

• Describe to what extent these type of general system context factors are relevant for your action lab.

## 5. PARTICIPATORY MONITORING

Participatory monitoring is an established and accepted way for the public to make informed decisions. Through the collection of data that is credible to multiple parties, participatory monitoring can become an essential instrument for generating trust. Participatory monitoring strategies in the action labs allow to evaluate baseline water quality conditions, identify pollution pathways, and design appropriate measures. By engaging and combining monitoring data from different actors in the action labs into harmonized datasets we create trust, knowledge, and awareness among actors.[31] Monitoring is usually conducted by outsiders (research organizations, environmental agencies, and drinking water companies) to local farmers and consumers, thus creating an information gap between farmers and monitoring agencies. Moreover, current groundwater and surface water sampling are designed for status and trend assessment of water bodies but less focused on assessing the effectiveness of the measures introduced to prevent or limit the inputs of pollutants.[31,32] To increase the credibility of the monitoring, farmers and consumer groups need to be engaged in the design and the setup of water monitoring, as well as in the results of the monitoring through the appropriate ICT tools and monitoring apps. Dedicated, high-frequent and innovative monitoring using test kits,[33] water quality sensors and continuous water sampling,[34,35] and analysis techniques[36] allows to relate measured concentrations to input pathways in the water system. In some cases multiresidue pesticides analysis specific for culture, as, for example, for vineyards, may be needed in order to identify the contribution of that culture to general pollution of the water system. Other innovative monitoring strategies include the isotopic analysis of nitrogen species[37,38] as a way to identify and distinguish between manure and inorganic amendment, the application of infrared imaging to detect illegal discharges of sewage within a catchment,[39] the performance of a vertical profiling of nitrogen species concentrations in groundwater[40] to evaluate the fate of nitrates, and soil sampling to verify correctness of fertilizer's management plans, monitoring of benthic bioindicators of surface water (algae) and use of water kits for self-monitoring of water nutrients as part of information and awareness campaigns. The participatory monitoring also lays down the basis for indicators related to water quality status, complementary to existing water quality status indicators.

## 6. BEST MANAGEMENT PRACTICES

To protect drinking water sources and to improve the water quality, a mix of cost-effective measures (sustainable use of chemicals, management practices, mitigation measures, point source pollution treatment) needs to be established at the catchment or action lab scale with a maximum impact on improvement of (raw) water quality. The farmers could be considered as producers not only of safe food but also of good water quality (delivery of green and blue services) allowing, e.g., the drinking or bottled water industry to produce water without need for extensive purification and treatment. Extensive collective work has been done on the dissemination of measures at the plot or field scale (e.g., TOPPS-Life+[41] and MAgPie[42] projects) but not yet on the evaluation of the effect of measures at the catchment scale, i.e., dependent on the location in the catchment or groundwater capturing area. This challenge by applying a farming systems design approach combining mitigation measures at the landscape level. Innovative farming systems (e.g., interridge bunding, conservation agriculture, precision agriculture, IPM) and mitigation measures and combinations of measures[43–46] (e.g., tests with biopurification systems for farmyard wastewater, washing/cleaning place for the sprayers, drift–reducing nozzles, land–use measures) are needed to provide good water quality at the catchment level under different pedoclimatic, cultural, economic, and geographic conditions.

## 7. COLLABORATIVE SOFTWARE TOOLS

Collaborative or participatory software tools can assist the multiactor process by creating a transparent, consistent, coowned environment in which the landscape information, the water quality status, and the intended actions can be visualized. An example of such a participatory tool is the Agricultural Land Information System (ALIS).[47] ALIS is a map-based tool developed to support farmland preservation decisions. It fulfills key factors for tool implementation: compatibility with the intended decision task, flexibility, transparency, explicit recognition of the need to not entirely rely on maps, ownership, trust, and ability to manage future tool improvements. An important ambition and innovation in multiactor approaches is the integration of the biophysical system knowledge in a collaborative management tool which enables transparent communication and interaction with farmers, local actors, and stakeholders. In different countries harmonized platforms[48,49] already exist for web-based visualization of water quality data, and geospatial information with user interaction (easy access to water quality data and geospatial information). Existing tools[50] can be used to identify optimal sets of mitigation methods at farm level, but they do not take into account the effect of the position and spatial layout of the fields in the catchment or the groundwater capturing area surrounding a drinking water well. Another example of a tool, specifically for water quality management, is the WaterProtect tool, which is built on existing components for water quality and landscape visualization and selection of cost-efficient measures for water quality at the catchment scale.[51] It is an interactive collaborative software platform, able to visualize monitoring results (monitoring app), landscape information, risk areas, and scenarios of implementation of mitigation measures.

# 8. RESULTS FROM EXAMPLE CASE STUDIES
## 8.1 Cicindria Case Study

The Cicindria study catchment (1075 ha) is a small agricultural catchment located in the east of Belgium in a fruit growing region. It is characterized by mainly agricultural land use with some residential land use and in the headwater frequently high concentrations of pesticides have been detected. Pesticide concentrations in rivers generally have a very dynamic signature and are strongly dependent on time and space. The dynamic time course is due to the time- and space-variant input conditions resulting from fast overland (runoff and erosion, direct losses) and subsurface flow (artificial drainage), directly connecting surfaces and/or agricultural fields where pesticides are applied, to receiving rivers. In order to increase the effectiveness of mitigation measures a thorough understanding of pesticide behavior at the watershed scale and effective communication to farmers is needed. Stakeholders in this process are farmers and farmers advisory (PCFruit npo, PIBO campus), research organizations (VITO), plant protection industry (GESSG), land managers (VLM), water managers (Water Board

Sint-Truiden), the municipalities of Sint-Truiden and Gingelom, the utilities Aquafin and Infrax, and the environment agency (VMM).

In the Cicindria case a targeted approach was developed where first a map was derived with priority zones for applying mitigation measures. This information in combination with information from a pesticide monitoring campaign was used to communicate to local farmers with focus on those farmers with potentially a significant impact on the pesticide load to the river.

A risk map representing the risk of pesticide runoff to the river was derived based on information about the topography, crop cover, the estimated pesticide use, the potential erosion risk, and the connectivity of the agricultural parcels to the river. Subsequently, the theoretical risk map was validated in the field using field observations of runoff during stormflow events, and using observations of roads short-circuiting the runoff to the river. The risk map in combination with monitoring results of pesticide concentrations in the river was further used to communicate to the local farmers.

From the validated risk map (Fig. 7) priority zones were defined for measures related to erosion control. The information was used to target farmers that may have a significant impact on the pesticide load to surface water. Those farmers were encouraged to participate in a voluntary erosion control program supported by the local government, starting from 2016 on. A strategy based on the installation of grassed zones located at the lower end of the cultivated fields (vegetated bufferstrips), in the major runoff pathways (i.e., grassed waterways), and the creation of (small) sediment depositions sites within the cultivated catchment is potentially very effective and efficient.

Communication was done in two steps[1]: general information meetings to the local community of farmers and[2] visits to farmers with fields located in the identified priority zones. During the first year of active communication and involvement of farmers, 11 grass buffer strips have been installed in the catchment with a width of 9 or 21 m covering a total area of 8.46 ha. The effect of the mitigation measures on water quality is further assessed in a monitoring campaign.

## 8.2 Bollaertbeek Case Study

The Bollaertbeek catchment (23 km$^2$, 160 farmers) with mixed urban/rural land use is part of the surface water capturing area of the drinking water

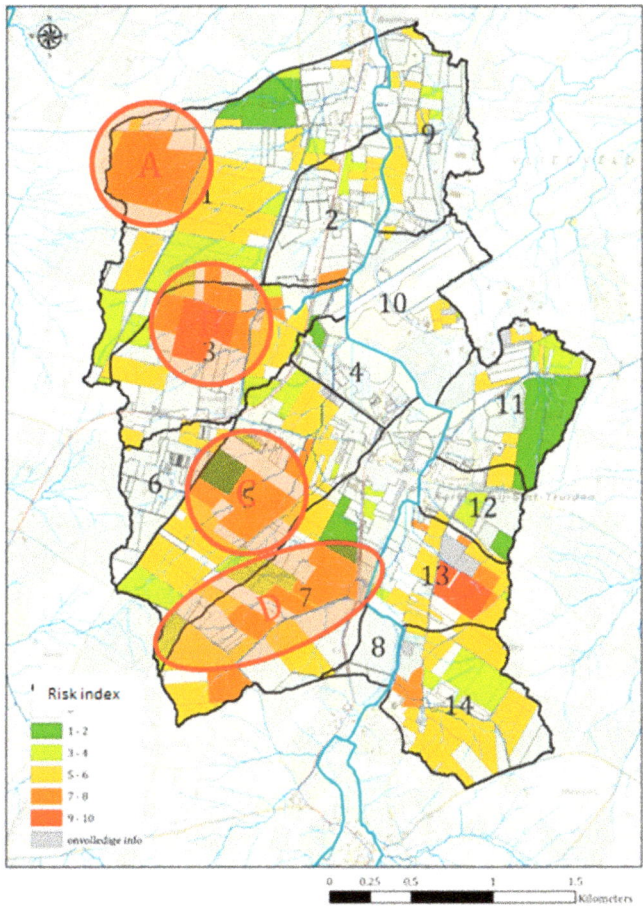

**Fig. 7** Validated pesticide runoff risk map for the Cicindria catchment and priority zones A, B, C, and D.

production company De Watergroep. PPP concentrations are a problem for water quality in the catchment. Despite several extension and training campaigns on water contamination problems by PPPs, there still is a low uptake of mitigation measures. Nitrate concentrations in surface water in the region have improved after the establishment of water quality groups (by CVBB). The CVBB approach is a combination of monitoring, setting up discussion groups, advising farmers on fertilization and fertilization plans. A new governance strategy to decrease PPP concentrations in surface water in the region is being developed building on the experience from the water quality groups (CVBB) and the lessons learnt from other successful cases of increased uptake of BMPs of crop protection in the nearby Kemmelbeek catchment

and the Cicindria catchment (see later). The interaction with the local actors is organized in multiple multiactor groups. The first multiactor group is characterized by:
- Large group of all stakeholders.
- Lower level of participation.
- Possibility to be informed, newsletter, etc.

The subsequent multiactor groups focus on specific aspects related to water quality and the ways to improve the water quality. It is characterized by:
- Intensive working group.
- All stakeholders have the possibility to join this multiactor group.
- Collaborative discussion and selection of possible actions to improve water quality.
- Implementation of actions.

The multiactor moments are carefully prepared. Stakeholders were motivated to participate through direct e-mail (twice), letter, and phone. Interviews with some stakeholders before the meeting were done. The first multiactor meeting aimed to describe the water quality problem to the farmers in the action lab and discussions about possible solutions (What idea did you have about the water quality regarding PPP in the area of the Bollaertbeek before you came to the meeting? How important is the water quality regarding PPP in the area of the Bollaertbeek?). The actors invited were the farmers practice center (Inagro), research organizations (VITO and ILVO), the Flemish Environment Agency (VMM), the water producer De Watergroep, the association of PPP producers (Phytofar), the municipalities (Ieper and Heuvelland), farmers inside and outside the action lab, and trade unions. Methods used were presentation and discussion groups managed by the practice centre and the research organizations and moderated by sociologists. The multiactor group quickly found a common objective to realize a better water quality of the Bollaertbeek. Major challenges faced with were the identification of relevant stakeholders, language issues, time constraints to organize and to present, selecting the right channel for communication. It was a new concept for such a meeting in the region, farmers are not used to discuss much in group. The main advantages or positive results of organizing this session were that farmers and stakeholders were very positive about the meeting, and about the concept of the meeting. The farmers present in the meeting were discussing well and searching for positive solutions for all actors. The actor meeting learned about possible BMPs to implement, and the "positive acting farmers."

An important instrument to facilitate actor discussions in the Bollaertbeek catchment is the WaterProtect tool. Actors expressed their needs to the tool:

• A user-friendly web-based collaborative management tool with collective participation of farmers and consumers.
• The tool provides knowledge on the functioning of the water system through the integration of geospatial information (landscape, soils, geology, land use, crops, urban use, rivers, water collection), water monitoring, results of hydrological and hydrogeological modeling of the interaction between farming systems and water quality, and cost-efficiency of mitigation measures.
• The tool shall use transparent and understandable indicators to ensure reliable and comparable data in order to involve farmers and citizens.
• The tool shall support the identification and localization of measures to be implemented.
• The user needs only a browser to consult the WaterProtect tool.
• The web-based tool is to facilitate the exchange of knowledge among actors and explore the optimal local solutions.
• The tool should organize and visualize data of the use case.
• The tool should cover functionalities to be used in a collaborative decision-making process.
• The tool should include a module for identifying vulnerable and risk zones for water pollution.
• The tool may include new modules for making predictions (what is the effect of mitigation on water quality?), assess risk areas, effects of mitigation measures, and the cost-effectiveness of measures.
• The tool is based on an open source GIS environment.
• They should organize data in a digestible manner so it can reach various target groups in the action labs (farmers and decision makers).

The benefits of the tool are:

• Creating transparency in the connection between farming systems and water quality will result in more transparent relationships between the application of nitrate and pesticides (and their relevant metabolites) and their occurrence in drinking water intake.
• Increasing local engagement by sharing data from participatory monitoring.
• Reporting tool for water quality in relation to land use and level of nitrate pollution of drinking and surface water in catchments.

- Facilitating the use and interpretation of collected monitoring data and converting technical data into easily readable/digestible information.

An overview of a number of current functionalities of the tool is shown in Fig. 8. The tool is continuously adapted to specific user needs.

## 8.3 Vittel Case Study

The Vittel case study concerns an agreement between a private company and the farmers. In 1988, the company Vittel detected increasing nitrate concentrations in aquifers due to intensive agriculture in the catchment area of the Vittel wells. Multiple alternative strategies were considered: doing nothing, drilling new wells, buying land in the catchment area, law suits against the farmers, or making contracts with the farmers. The first four alternatives were considered not cost-efficient and expensive, and some alternatives had severe consequences (the term "mineral water" is strictly regulated in France). The last alternative, making contracts with farmers, seemed most feasible since the number of farmers was limited and the negotiation procedure was considered feasible.[52] In this process, the French research organization INRA played a crucial role in bringing the positions of farmers and company closer together. The company had no idea about what measures could be efficient to reduce the amounts of nitrate in the groundwater and what these measures would cost and how it would affect the income of the farmer. The farmers saw the company as a multinational that wanted agriculture being pushed out of the catchment. INRA acted as the intermediate in the negotiations.[52] The negotiations led to the establishment of contracts between the farmers and the company, in which the farmers agreed to have less intensive dairy farming and pasture management. The following measures were imposed to the farmers: no corn production, a ban on pesticides, a maximum of 1 livestock unit per hectare, balanced feed, and modernization of manure storage. Contracts had a duration of 18–30 years to ensure operations for the farmers. Farmers received as compensation for income reduction a funding of 200 € per hectare per year for a period of 5 years, support for the purchase of machinery up to 150,000 € per farm, service provision in the form of manure, and compost spreading on the fields and free technical support. The compensation varied among farmers depending on the distance of the farming practices to the wells and the total amount of land of the farmer in the catchment.[52] The negotiation process and the compromise took a long time since farms were heterogeneous

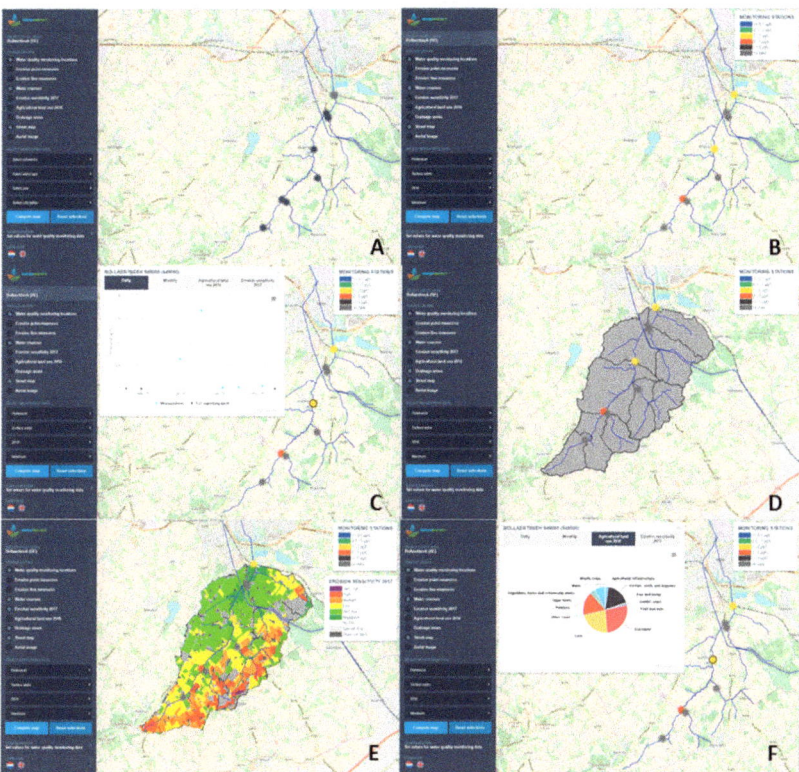

**Fig. 8** WaterProtect webtool screenshots. (A) In the first screen the area of the catchment (Bollaertbeek) is shown with the watercourses in the area and the location of the water quality monitoring stations. On the *left* a selection bar is shown where the user can select different maps to be shown (Water stations: location of water quality monitoring stations; Rivers: watercourses; Land use: agricultural land use in the area; Erosion sensitivity: potential erosion for the agricultural parcels; Drainage areas: drainage areas connected to the monitoring stations). (B) Apart from different base maps also some selections need to be made for the visualization of water quality data: Choose substance: selection of parameter (plant protection product or nutrient); Choose source: surface water or groundwater; Choose year: year of monitoring; Choose standard: which evaluation: average concentration or maximum concentration. When the user clicks on "Compute Map," the monitoring stations will be colored according to the evaluation against an environmental quality standard. (C) Additionally the user can click on a monitoring station and a pop-up appears with the tabs "Daily," "Monthly," "Land use," and "Erosion sensitivity." The tab "Daily" displays the individual measurements as a time series. The tab "Monthly" gives for every month the average and maximum value as well as the yearly average. (D) "Drainage areas" is a map of the catchment and subcatchments related to each monitoring station and can be selected in combination with any of the other maps. (E) "Erosion sensitivity" is the map of potential erosion for each agricultural field and cannot be selected in combination with "Land use"; when this map is selected, the "Land use" map is automatically deselected. (F) By clicking a monitoring station the statistics of "Land use" and "Erosion sensitivity" of the catchment associated to the monitoring station are shown as pie charts.

and therefore the determination of costs for the measures and the height of the compensation were a tedious job.[53] Farmers were subdivided into four groups, each with an own strategy, objectives and constraints, and their willingness and capacity to perform the changes: small farms (on average 19 ha), specialized dairy farms (less than 50 ha), farms with less than 135 ha, and farms larger than 135 ha. One of the major points of discussion was the basis on which the compensations needed to be calculated: the costs for the farmers or the benefits of the company. The compensation fee could not be lower than the opportunity cost of the proposed measures (loss of production value or loss of income, investments, training and education cost, etc.) plus an extrastimulus to have farmers changing their behavior. The maximum value of the compensation fee was the opportunity cost for the company or the increase in value of the bottled water due to the guarantee of the quality of the bottled water (avoided market loss), disregarding profit. At the start of the negotiations, the company wanted the minimum option, while the farmers wanted the maximum option. The farmers were in a good negotiation position since they could assess the costs very well as well as the impact of the proposed measures on their farm. Farmers at strategic locations (close to the well) had a great power (monopoly power) in the negotiation process because their participation was necessary (nonreplaceable) and since they had a lot of land in the catchment. This led to opportunistic behavior resulting in elevated transaction costs. Some farming organizations discouraged farmers to enter into the negotiations. On the other hand, the company delivered a lot of jobs in a region with high unemployment. This helped in reaching a compromise between the farmers and the company. The average fee for the farmers was 200 € per hectare per year and the cost for the company amounted to 980 € per hectare per year or 1.52 euro per cubic meter bottled water for the first 7 years. The total cost for the company over the 7-year period amounted to 24.25 million euro of which 47% went to compensations for the farmers, 38% went to buying the land (which were rented to other farmers), and 15% for investment in new farming machinery. In the Vittel case there was 1 initiator (the company) and 26 providers (farms). The contracts are voluntary and differentiated according to the cost structure and location of the various farms. There is a scientifically established link between ecosystem services (water infiltration and soil nitrate levels) and the management practices. The fees are intended to new investments and cost recovery related to implementation of new measures and not directly to reduce nitrate levels in the aquifers.

# 9. CONCLUSIONS

We think a multiactor approach as described here will be instrumental to achieve good water quality in intensively cultivated landscapes. Farmers need to be made more aware about the effects of their behavior and much more engaged in the process of water quality management in a catchment. Engagement can be stimulated by taking the farmers into a well-thought governance process that takes into account a multitude of options and financing regimes allowing negotiation between actors to be more successful. Besides good governance, also engagement of farmers and actors in the water quality monitoring through codesign, common implementation and follow-up of the monitoring is crucial. Long lists and practice sheets of good agricultural practices and mitigation measures are already available for a long time but seem not to be leading to real implementation because of lack of engagement. Transparency in the relationship between farmer behavior and water quality is indispensable to convince farmers to change their behavior. Mitigation measures should take the specific biophysical characteristics of the landscape into account to prove their cost-effectiveness and adoption by the farmer. Collaborative tools are a great help in making both water quality status and effect of mitigation measures and management practices on the water quality clear and transparent. Bringing information close to the farmer using ICT tools will help to lower the barrier for implementation of the measures and real sustainable use of chemicals in agriculture.

## ACKNOWLEDGMENTS

Parts of this work were financed from the European Union's Horizon 2020 Research and Innovation Programme under Grant agreement 717450.

## REFERENCES

1. European Commission. *Report from the commission to the European Parliament and the council on the implementation of the water framework directive (2000/60/EC) river basin management plans.* 2012.
2. European Commission. *Eurostat statistics explained—agri-environmental indicator—nitrate pollution of water,* European Commission; 2012.
3. European Environment Agency. *European waters—current status and future challenges.* European Environment Agency; 2012EEA Report No. 9.
4. European Environment Agency. *The European environment—state and outlook 2015: an integrated assessment of the European Environment.* Copenhagen: European Environment Agency; 2015.

5. OECD. Water governance in OECD countries: a multi-level approach. In: *OECD studies on water*. OECD Publishing; 2011.

6. Ostrom E. A diagnostic approach for going beyond panaceas. *Proc Natl Acad Sci USA* 2007;**104**(39):15181–7.

7. Ostrom E. A general framework for analyzing the sustainability of socio-ecological systems. *Science* 2009;**325**(5939):419–22.

8. OECD. *Stakeholder engagement for inclusive water governance*. In: *OECD studies on water*. Paris: OECD Publishing; 2015. https://doi.org/10.1787/9789264231122-en.

9. Rogers P, Hall A. Effective water governance. In: *TEC background papers No. 7*. Stockholm: Global Water Partnership; 2002.

10. Rogge E, Dessein J, Verhoeve A. The organisation of complexity: a set of five components to organise the social interface of rural policy making. *Land Use Policy* 2013;**35**:329–40.

11. Emerson K, Nabatchi T, Balogh S. An integrative framework for collaborative governance. *J Public Adm Res Theory* 2011;**22**(1):1–29.

12. Pahl-Wostl C. A conceptual framework for analysing adaptive capacity and multi-level learning processes in resource governance regimes. *Glob Environ Chang* 2009;**19**(3): 354–65.

13. Baxter J, Eyles J. Evaluating qualitative research in social geography: establishing 'rigour' in interview analysis. *Trans Inst Br Geogr* 1997;**22**:505–25.

14. Morgan DL. *The focus group guidebook*. SAGE Publications Inc; 1997.

15. Creswell J. *Research design: qualitative, quantitative and mixed method approaches*. Thousand Oaks: Sage; 2003.

16. Patton M. *Qualitative research and evaluation methods*. Thousand Oaks: Sage; 2002.

17. Macken-Walsh A, O'Dwyer T. Discussion groups: five key ingredients for success. *Irish Farmers' J* 2016;7th April, 2016.

18. Messely L, Rogge E, Dessein J. Using the rural web in dialogue with regional stakeholders. *J Rural Stud* 2013;**32**:400–10.

19. Tropp. Water governance: trends and needs for new capacity development. *Water Policy* 2007;**9**(S2):19–30. WWAP. 2006. Water, a shared responsibility. The United Nations World Water Development report 2, UNESCO Publishing, 2006.

20. Heley. Soft spaces, fuzzy boundaries and spatial governance in post-devolution wales. *Int J Urban Reg Res* 2013;**37**(4):1325–48.

21. Carpenter S, Folke C, Norstrom A, Olsson O, Schultz L, Agarwal B. Program on ecosystem change and society: an international research strategy for integrated social-ecological systems. *Curr Opin Environ Sustain* 2012;**4**:134–8.

22. North DC. *Institutions, institutional change and economic performance*. Cambridge: Cambridge University Press; 1990.

23. GWP. Integrated water resources management (IWRM) and water efficiency plans by 2005. In: *TAC background papers no. 10*. Stockholm, Sweden: GWP. Global Water Partnership—Technical Advisory Committee (GWP-TAC); 2004.

24. Hill M. A starting point: understanding governance, good governance and water governance. In: Climate change and water governance: adaptive capacity in Chile and Switzerland. *Adv Glob Change Res*, vol. 54. Netherlands: Springer; 2013. ISBN: 978-94-007-5796-7. p. 17–26. https://doi.org/10.1007/978-94-007-5796-7.

25. Bardach E. *Getting agencies to work together*. Washington, CD: Brookings Institution; 1998.

26. Ring PS, Van de Ven AH. Developmental processes of cooperative inter-organizational relationships. *Acad Manage Rev* 1994;**19**:90–118.

27. Thomson A, Perry J. Collaboration processes: inside the black box. *Public Adm Rev* 2006;**66**:20–32.

28. Fisher R, Brown S. *Getting together: building relationships as we negotiate.* New York, NY: Penguin Books; 1989.
29. Bryson JM, Crosby BC, Middleton Stone M. The design and implementation of cross-sector collaborations: propositions form the literature. *Public Adm Rev* 2006;**66**: 44–55.
30. Selin S, Chavez D. Developing a collaborative model for environmental planning and management. *Environ Manag* 1995;**19**:189–95.
31. http://www.fao.org/docrep/x5307e/x5307e05.htm.
32. EC, Common Implementation Strategy for the Water Framework Directive (2000/60/ EC). Guidance document No. 15. Guidance on groundwater monitoring, European Commission, 2007.
33. http://www.action4acomb.co.uk/wp-content/uploads/2015/03/Red-Burn_Training Card_WaterQuality-v0.1.pdf.
34. Mellander P-E, Jordan P, Shore M, Melland AR, Shortle G. Flow paths and phosphorus transfer pathways in two agricultural streams with contrasting flow controls. *Hydrol Process* 2015;**29**:3504–18.
35. Mellander P-E, Melland AR, Jordan P, Wall DP, Murphy PNC, Shortle G. Quantifying nutrient transfer pathways in agricultural catchments using high temporal resolution data. *Environ Sci Pol* 2012;**24**:44–57.
36. Postigo C, López de Alda MJ, Barceló D, Ginebreda A, Garrido T, Fraile J. Analysis and occurrence of selected medium to highly polar pesticides in groundwater of Catalonia (NE Spain): an approach based on on-line solid phase extraction-liquid chromatography-electrospray-tandem mass spectrometry detection. *J Hydrol* 2010;**383**:83–92.
37. Puig R, et al. Multi-isotopic study (15N, 34S, 18O, 13C) to identify processes affecting nitrate and sulfate in response to local and regional groundwater mixing in a large-scale flow system. *Appl Geochem* 2013;**32**:129–41.
38. Urresti-Estala B, et al. Application of stable isotopes (δ34S-SO4, δ18O-SO4, δ15N-NO3, δ18O-NO3) to determine natural background and contamination sources in the Guadalhorce River basin (southern Spain). *Sci Total Environ* 2015;**506–507**: 46–57.
39. M. Lega and RMA Napoli "Aerial infrared thermography in the surface waters contamination monitoring" Proceedings of the 11th international conference on environmental science and technology. Chania, Crete, Greece, 3-5/9/2009.
40. Bourke, et al. Comparison of continuous core profiles and monitoring wells for assessing groundwater contamination by agricultural nitrate. *Groundwater Monit Remediat* 2015; **35**:110–7.
41. http://www.topps-life.org/.
42. Alix A, Brown C, Capri E, Goerlitz G, Golla B, Knauer K, Laabs V, Mackay N, Marchis M, Poulsen V, Alonso Prados E, Reinert W and Streloke M, 2017. Mitigating the risks of plant protection products in the environment. SETAC editions. ISBN: 978-1-880611-99-9, Publication Date: May 2017; Publisher: SETAC. https://www.setac.org/store/ViewProduct.aspx?id=9006489
43. De Wilde T, Spanoghe P. Review of on-farm bioremediation systems to reduce the occurrence of point source contamination. *Pest Manag Sci* 2007;**63**:111–28.
44. Felsot AS, Unsworth JB, et al. Agrochemical spray drift; assessment and mitigation—a review. *J Environ Sci Health* 2010;**4**(1):1–24.
45. Boye K, Jarvis N, et al. *Pesticide run-off to Swedish surface waters and appropriate mitigation strategies.* Swedish University of Agricultural Science; 2012. research report, 47 pages.
46. Reichenberger S, Bach M. Mitigation strategies to reduce pesticide inputs into ground- and surface water and their effectiveness: a review. *Sci Total Environ* 2007;**384**: 1–35.

47. Kerselaers E, Rogge E, Lauwers L, Van Huylenbroeck G. Decision support for prioritising of land to be preserved for agriculture: can participatory tool development help? *Comput Electron Agric* 2015;**110**:208–20.

48. OpenTEA webplatform for e-learning in sustainable use of plant protection products: www.opentea.eu

49. WaterProtect BE webviewer for surface water and groundwater quality related to pesticides: www.water-protect.be

50. Gooday R, Anthony S. *Mitigation method-centric framework for evaluating cost-effectiveness Defra project WQ0106 (Module 3)*. Wolverhampton: ADAS UK Ltd; March 2010.

51. Broekx S. *Modelling tools for cost-effective water management*. PhD Dissertation, University of Ghent; 2014.

52. Déprés C, Grolleau G, Mzoughi N. In: *Contracting for environmental property rights: the case of Vittel. Paper prepared for presentation at the 11th congress of the EAAE (European Association of Agricultural Economists), 'The Future of Rural Europe in the Global Agri-Food System'*; 2005. 24–27 augustus 2005, Kopenhagen.

53. Perrot-Maître D. *The Vittel payments for ecosystem services: a perfect PES-case?* London, UK: International Institute for Environmental and Development; 2006.

CHAPTER THREE

# Certification and Added Value for Farm Productions

**Pieter Ravaglia\*,1, Jacopo Famiglietti†, Fiamma Valentino‡**

\*Department for Sustainable food process, Università Cattolica del Sacro Cuore, Piacenza, Italy
†Independent author
‡DG for Sustainable Development (SVI), Italian Ministry for the Environment Land and Sea, Rome, Italy
1Corresponding author: e-mail address: pieter.ravaglia@unicatt.it

## Contents

| | |
|---|---|
| 1. Introduction | 65 |
| 2. Agriculture in Europe | 66 |
| 3. Quality in the Agro-Food Sector | 68 |
| 4. Origin of Sustainable Consumption | 69 |
| 5. 360° Quality: Classification of Public and Private Certification Initiatives | 72 |
|     5.1 Classification of Public Initiatives | 73 |
|     5.2 Classification of Private Initiatives | 85 |
| 6. Focus on Geographical Indications | 91 |
| 7. Multiplication of Environmental and Sustainability Labels | 93 |
| 8. European Consumer Behavior | 94 |
| 9. Conclusions and Future Developments | 95 |
| Annex I | 98 |
| Annex II | 101 |
| References | 106 |

## Abstract

Since the 1990s, a large number of companies in the agricultural sector have decided to rely on quality, a challenge linked to certification systems, especially those of Protected Geographic Indication (PGI), Protected Designation of Origin (PDO), and Organic production. Since then, the public and private certifications linked to the concept of quality productions have been the engine of an economic revolution that has transformed the "typical product" from a niche sector to a sector that has driven agricultural innovation. This sector represents today a significant part of the European economy considered successful by the rest of the world. EU agro-food products, both as agricultural raw materials and final products, enjoy now an undisputed and envied reputation of the highest quality on the world market, especially for Italian products.

At the same time, new patterns of food consumption and the economic crisis are also driving businesses to a new strategic repositioning. The strongest signals come from large-scale retailers and consumers who demand a real differentiation of production, with particular regard to their environmental and social impacts.

*Advances in Chemical Pollution, Environmental Management and Protection*, Volume 2
ISSN 2468-9289
http://dx.doi.org/10.1016/bs.apmp.2018.03.003
© 2018 Elsevier Inc.
All rights reserved.

This work deals with the issue of quality, providing an overview of the main quality programs active at European level with a more specific analysis of those that lead to an innovative approach to sustainability. "Quality innovation" is the challenge for the certification sector today, and sustainability aspects can be the centerpiece of this transformation.

**Keywords:** Sustainable consumption, Food quality schemes, Food sustainability schemes, Certification scheme, Quality innovation, Agro-food products, Certified schemes

## ABBREVIATIONS

| | |
|---|---|
| **AOC** | Appellation d'Origine Contrôlée |
| **B2B** | Business to Business |
| **B2C** | Business to Consumer |
| **BRC** | British Retail Consortium |
| **CAM** | Minimum Environmental Criteria |
| **CAP** | Common Agricultural Policy |
| **CPCC** | Control Points and Compliance Criteria |
| **DAC** | Districtus Austriae Controllatus |
| **DO** | Denominación de Origen |
| **DOC** | Denominazione di Origine Controllata |
| **DOCa** | Denominación de Origen Calificada |
| **DOCG** | Denominazione di Origine Controllata e Garantita |
| **EC** | European Commission |
| **EF** | Environmental Footprint |
| **EPD** | Environmental Product Declarations |
| **FMI** | Food Marketing Institute |
| **FSC** | Forest Stewardship Council |
| **GFSI** | Global Food Safety Initiative |
| **GHG** | greenhouse gas |
| **HACCP** | Hazard Analysis and Critical Control Poi |
| **HEV** | High Environmental Value |
| **IFA** | Integrated Farm Assurance |
| **IFOAM** | International Federation of Organic Agriculture Movements |
| **IGT** | Indicazione Geografica Tipica |
| **INAO** | Institut national de l'origine et de la qualité |
| **IPR** | Indicação de Proveniência Regulamentada |
| **LCA** | Life Cycle Assessment |
| **LCI** | Life Cycle Inventory analysis |
| **MGI** | Made Green in Italy |
| **OECD** | The Organization for Economic Cooperation and Development |
| **OEF** | Organization Environmental Footprint |
| **PDO** | Protected Designation of Origin |
| **PEF** | Product Environmental Footprint |
| **PEFC** | Programme for the Endorsement of Forest Certification Schemes |
| **PGI** | Protected Geographic Indication |

| **PRPs** | PRerequisite Programs |
|---|---|
| **QbA** | Qualitätswein bestimmter Anbaugebiete |
| **QmP** | Qualitätswein mit Prädikat |
| **QS** | Quality Systems |
| **QWPSR** | Quality Wine Produced in a Specific Region |
| **SFQ** | Safe Food Quality |
| **SMEs** | small- and medium-sized enterprises |
| **SQNPI** | National Integrated Production Quality System |
| **SQFI** | Safe Quality Food Institute |
| **TSG** | traditional speciality guaranteed |
| **UAA** | utilised agricultural area |
| **VDP** | Vin de Pays |
| **VR** | Vinho Regional |
| **VT** | Vino de la Tierra |
| **vzbv** | federation of German consumer organizations |

# 1. INTRODUCTION

Since the 1990s, a large number of companies in the agricultural sector have decided to rely on quality, a challenge linked to certification systems, especially those of Protected Geographic Indication (PGI), Protected Designation of Origin (PDO), and Organic production. Since then, the public and private certifications linked to the concept of quality productions have been the engine of an economic revolution that has transformed the "typical product" from a niche sector to a sector that has driven agricultural innovation. This sector represents today a significant part of the European economy considered successful by the rest of the world.

The evolution of regulations at European level has favored the development and diffusion of certified quality schemes, and currently, with the Product Environmental Footprint (PEF) pilot, also an environmental performance certification process will be launched this year. At the same time, new patterns of food consumption and the economic crisis are also driving businesses to a strategic repositioning on quality in all its forms (including sustainable quality).

The strongest signals come from large-scale retailers and consumers who demand a real differentiation of production, with regard to the quality and origin of agro-food production and to their environmental and social impacts that are becoming a major issue in public debates.

Over the past 20 years, the European system (particularly as regards Italy, France, Spain, and Portugal) has gained a strong reputation making certification a subject capable of directly involving agricultural producers, manufacturers, sales channels, and even consumers, who are increasingly aware of the quality, origin, and environmental impact of agro-food products.

In particular, the environmental issue is having a predominant influence on consumer choices, which also in the selection of products labeled PDO, PGI, and organic are conditioned by the image of territory protection and respect for social aspects that these brands carry with them.[1]

In this chapter, we will deal with the issue of quality also introducing the theme of sustainability and how now for farmers it is fundamental combine quality aspects at 360°. "Quality innovation" is the challenge for the certification sector today, and environmental aspects can be the centerpiece of this transformation.

## 2. AGRICULTURE IN EUROPE

The food and drink industry is the, EU-28, biggest manufacturing sector in terms of jobs and added value.[2] In 2013, there were 10.8 million agricultural holdings within the EU-28. The utilised agricultural area (UAA) in the EU-28 was almost 175 million hectares (some 40.0% of the total land area). In terms of UAA, France and Spain had the largest share of the EU-28's agricultural land, with 15.9% and 13.3% shares, respectively, followed by the United Kingdom and Germany with a share just under the 10%. By contrast, the largest number of agricultural farms was in Romania (3.6 million), where one-third (33.5%) of all the holdings in the EU-28 were located. Poland had the second highest share of agricultural holdings (13.2%), some way ahead of Italy (9.3%) and Spain (8.9%).[3] In Italy, the food industry is the second largest after metalworking, with a turnover of 132 billion euros in 2016.[4] From the last census in 2010, there were 1,620,884 active agricultural and zootechnical holdings with a total UAA of 12.9 million hectares.[5]

In the last decade, quality and environmental protection have become one of the major issues in the public debates.

The competitiveness of European agriculture and agro-food products is increasingly dependent on the quality of the products themselves, which is, as is now generally recognized, the only real weapon for dealing with the globalization of markets.

Today, the quality of products, in particular agro-food products, is a subject on which the attention and concerns of consumers are strongly focused, urging at all levels the taking of initiatives for protection, information, and promotion. The "Global Health and Ingredient-Sentiment" survey[6] showed that "Consumers want to eat in ways that address real dietary concerns, but they can't do it alone. They need help from food manufacturers to offer products formulated with an eye towards food sensitivities and other specialized diets, and they need help from retailers to stock shelves with a proper assortment of foods that cater to a wider variety of consumer needs," said Andrew Mandzy, Director, Strategic Insights, Nielsen. "This is a significant opportunity for food retailers and manufacturers, but even within individual markets, health and wellness is not a one-size fits all approach. Retailers and manufacturers need to identify high-potential segments and the drivers of engagement for these consumers and, then tailor their messages and products accordingly."[7]

The environmental impact of agriculture is a relevant issue. The European Commission (EC) estimates that agriculture and animal husbandry activities emissions in the EU-28 amount to 470.6 million tons of $CO_2$ equivalent in 2012, representing 10.3% of total greenhouse gas (GHG) emissions in the European Union (EU).[8]

The Paris Climate Conference (COP21) endorsed the principle that "climate change is an urgent and potentially irreversible threat to human societies and the planet." It therefore requires "maximum cooperation from all countries" with a view to "speeding up the reduction of greenhouse gas emissions."

In the recent Bonn Climate Conference (COP23), which ended on 18 November 2017 and preceded the COP24 in Katowice, the President of the European Committee of the Regions (CoR), Karl-Heinz Lambertz, stated that "Innovative approaches must be adopted at local level," and the challenge is "to change something in people's minds and behaviour."[9]

The European Commission is developing policies in this direction.

In the COMMUNICATION FROM THE COMMISSION TO THE EUROPEAN PARLIAMENT, THE COUNCIL, THE EUROPEAN ECONOMIC AND SOCIAL COMMITTEE AND THE COMMITTEE OF THE REGIONS—Roadmap to a Resource Efficient Europe,[10] the European Commission has set an important objective: by 2020, citizens and public authorities will be adequately encouraged to choose the most resource-efficient products and services, with correct price signals and clear environmental information. Their purchasing choices will incentivize companies to innovate and offer more efficient products (goods and services). Minimum environmental performance standards will be set to remove less efficient and more polluting products from the market. There will be strong consumer demand for more sustainable products and services.

In the COMMUNICATION FROM THE COMMISSION TO THE EUROPEAN PARLIAMENT AND THE COUNCIL Building the Single Market for Green Products—Facilitating better information on the environmental performance of products and organizations[11] the Commission announced that: "The Roadmap to a Resource Efficient Europe sets an ambitious target for 2020: to adequately encourage citizens and public authorities to choose the most resource-efficient products, through correct price signals and clear environmental information. It also recognizes the key role of the Internal Market in rewarding resource efficient products." This initiative "Building the Single Market for Green Products" is an important step in this direction.

This Communication sets out two methodologies for measuring the environmental performance of products and organizations and a set of principles on which to base their communication. The Communication is accompanied by a Commission Recommendation encouraging Member States and the private sector to use these new approaches, as appropriate, to improve the functioning of the internal market.

In order to achieve this, the Commission has set up the Single Market for Green Products initiative, for a shared European Environmental Footprint (EF) methodology.[12]

In the COUNCIL CONCLUSIONS ON ECO-INNOVATION: ENABLING THE TRANSITION TOWARDS A CIRCULAR ECONOMY the Council of the European Union invites the Commission "to explore also the possible uses of the Product Environmental Footprint (PEF) and Organisation Environmental Footprint (OEF) for measuring and communicating environmental information, taking full account of the need to maintain the competitiveness of member states."[13]

## 3. QUALITY IN THE AGRO-FOOD SECTOR

According to ISO 8402, quality is "the totality of characteristics of an entity that bear upon its ability to satisfy stated and implied needs."

From this definition, it is clear how quality is based on the figure of the consumer who, through the "product," satisfies his expectations. A human factor is intrinsic to quality assessment, which depends on the availability and knowledge of the product being consumed.

The good is seen as a "basket of attributes," however, in the case of food and agro-food products, not all characteristics are identifiable and therefore can be assessed in qualitative terms before actual consumption, and very often even after it. Following the Nelson classification[14] "search good" is defined as a product for which the verification of quality is possible before the purchase of the product itself, but this verification is expensive. On the other hand, "experience good" is defined as a product for which quality is only revealed after purchase, at the time of consumption. In all those cases in which, even after its purchase, the consumer is not able to determine the quality characteristics, it is defined "Credence good."[15]

The attributes of a food product range from "visual" characteristics (appearance, color, absence of external damage, etc.), "organoleptic" characteristics (bouquet, taste, etc.) and attributes that are not directly observable or verifiable by the consumer, often not even after consumption, the so-called Credence characteristics such as product salubrity, environmental and social aspects, and origin characteristics.

The consumer is more and more concerned about the attributes of "credence," quality in its broadest sense includes aspects such as health, environmental protection, animal welfare, aspects of origin, and protection of the territory, and embraces in an indissoluble way the concept of sustainability.

## 4. ORIGIN OF SUSTAINABLE CONSUMPTION

The discussion on sustainability evolved over 25 years ago, when the Brundtland Commission published their report "Our Common Future" in which it defines sustainable development as: "development that meets the needs of the present without compromising the ability of future generations to meet their own needs."[16]

The Earth Summit in Rio de Janeiro was followed by the World Summit on Sustainable Development in Johannesburg in 2002. Sustainable consumption and production was emphasized by the delegations participating at the World Summit as being essential for reaching sustainable development. A working definition of Sustainable Consumption was already provided at the "Oslo Symposium on Sustainable Consumption" in 1994, defining it in the following way: "Sustainable Consumption is the use of services and related products, which respond to basic needs and bring a better quality of life while minimising the use of natural resources and toxic materials as well as the emissions of waste and pollutants over the life cycle of the service or product so as not to jeopardise the needs of future generations."[17]

Another important outcome of the World Summit was the Johannesburg Plan of Implementation calling for a "10 Year Framework of Programmes" to be established that would promote Sustainable Consumption to achieve economic development without harming the environment.[18]

In 2012, 10 years after the World Summit in Johannesburg, the Rio + 20 United Nations Conference on Sustainable Development took place in Rio de Janeiro, and it has been a cornerstone of modern sustainable development policies and has strongly influenced the direction they have taken. It has enabled a consensus between the otherwise conflicting objectives of economic growth, social equity, and environmental protection by embracing the multidimensional concept of sustainable development. The Rio Declaration on Environment and Development, and Agenda 21 were the major outcome documents of the Rio conference. Agenda 21 laid out specific actions for integrating and attaining social, economic, and environmental objectives, including the role of major groups of stakeholders. The conference has been conceived as a landmark event in the global movement for sustainable development. As the main outcome, world leaders decided to launch a process for the development of a set of Sustainable Development Goals (SDGs), which will constitute the goals of the 2030 agenda for sustainable development, thus replacing the Millennium Development Goals after 2015.[19]

For many years the concept of sustainability has been used as a synonym for environmental sustainability. In the 20th century, the concept was broadened to include an economic and a social perspective. Those three perspectives have since then become a common approach for evaluating the progress toward sustainability.[20]

The three pillars of sustainability can be described as follows:
- *Environmental sustainability*: Environmental sustainability is a condition of balance, resilience, and interconnectedness that allows human society to satisfy its needs while neither exceeding the capacity of its supporting ecosystems to continue to regenerate the services necessary to meet those needs nor by our actions diminishing biological diversity.[21]
- *Economic sustainability*: the capacity of an economic system to generate a constant and improving growth of its economic indicators. In particular, the capacity to generate incomes and employment in order to sustain the populations. Within a territorial system, economic sustainability means

the capability, through the most efficient mix of resources, to produce and maintain the highest added value, in order to enhance the specificity of territorial products and services.[22]

- *Social sustainability*: the ability to guarantee welfare (security, health, and education), equitably distributed among social classes and gender. Within a territory, Social Sustainability means the capacity of the different social actors (stakeholders), to interact efficiently, to aim toward the same goals, encouraged by the close interaction of the institutions, at all levels.[22]

In 1995 UNESCO and then the World Summit on Sustainable Development have asked for the inclusion of *Culture* in the sustainable development model, since culture ultimately shapes what we mean by development and determines how people act in the world. This new approach addresses the relation between culture and sustainable development through dual means: first, the development of the cultural sector itself (i.e., heritage, creativity, cultural industries, crafts, and cultural tourism); and second, ensuring that culture has its rightful place in all public policies, particularly those related to education, the economy, science, communication, environment, social cohesion, and international cooperation.

Different authors have appealed to the use of images to increase the understanding of complex concepts, as sustainability and his three dimensions.[23] One of the most used representation is a stool with three legs. By inserting the new cultural pillar the authors likes to imagine it as a fundamental element where the weight of the three pillars is borne by culture, because sustainability does not exist without awareness (Fig. 1).

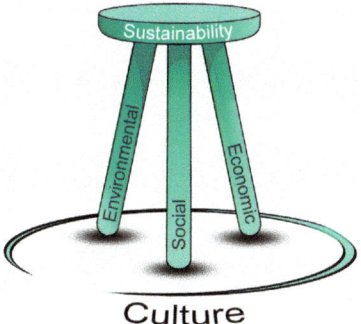

**Fig. 1** Sustainability based on culture. *Modified from EEA. 2002. Sustainable development reporting: Frameworks and aggregated indicators. Unpublished report of the EEA Expert Group on SoE Guidelines and Reporting. Copenhagen: European Environment Agency.*

## 5. 360° QUALITY: CLASSIFICATION OF PUBLIC AND PRIVATE CERTIFICATION INITIATIVES

As mentioned earlier, the products are characterized by "visual," "organoleptic," and "credence" characteristics. Some of these attributes/characteristics are "essential" and concern the fulfillment of collective interests such as trade requirements related to health and consumer information. The noncompliance with these requirements results in a ban on the product being marketed. Other attributes, instead, are "additional" and concern the guarantee of product quality principles, help to protect the reputation of companies, and in some cases may form the basis for obtaining a competitive advantage.

The "essential" attributes are often regulated by the public authority, which intervenes through different regulatory models (e.g., mandatory standards, certification schemes at regional or national level), which aim to achieve optimal quality levels in order to ensure consumer protection. Outside this area where public authorities are obliged to legislate, voluntary public and private regulation fill the gap created by "additional" consumer expectations. Those regulations use the instrument of third-party certification to guarantee to the consumer the respect of production standards or higher quality levels.

In 2012, the European Union issued EU Regulation No. 1151[24] "Quality package" on the quality of agricultural products, which proposes, in particular, the adoption of a common simplified and shortened registration procedure for geographical indications and traditional specialities, as well as clearer provisions on the relationship between trademarks and geographical indications, on the role of applicant associations, on the definition of "traditional speciality guaranteed" (TSG) and on the increasing demand for consumer information on "voluntary quality claims" such as the indication "Mountain Product." The new provisions establish the rules for agricultural products and foodstuffs, but exclude from their scope the disciplines on geographical indications relating to wines, spirits, and aromatized wines disciplined by the (EU) Regulation No. 1308/2013.

The Regulation No. 1151 is linked to the Communication on the Common Agricultural Policy (CAP) toward 2020, to the priorities set out in the Europe 2020 Communication and to the guiding principles of consumer information policy.

The European Package was created to protect agro-food quality by setting standards that affect various areas: from health care to information on the characteristics of the product (labeling) and its origin. In particular, the European provisions concern the protection of products with quality characteristics deriving from the specific production area (PDO), the typical production tradition of a specific locality (PGI), and the protection of intellectual property rights. This has enabled to tune and regulate the use of quality labels, which above all in their Business to Consumer (B2C) connotations were creating confusion among consumers by threatening the transparency of information, especially as regards the credibility of claims, the transparency of schemes and the excessive increase in costs for small producers by excluding or not promoting them in market transactions.

Later is an attempt, although certainly not in an exhaustive way, to classify the initiatives for quality "protection" of product by dividing them into public and private initiatives. The classification reported here also includes all those "sustainable" or "environmental" quality initiatives that are being developed at European level and which are now unavoidably linked to the concept of "360° quality."

A summary table of the public and private initiatives described in this chapter is provided in Annex II.

## 5.1 Classification of Public Initiatives

Public initiatives can be classified into two different subgroups: compulsory and voluntary.

### 5.1.1 Compulsory EU Public Initiatives

The public authority has always been the guarantor of the food market and has tried to preserve its proper functioning. The public authority, especially in the agro-food sector, has a higher degree of "responsibility" than the other productive sectors, both in terms of costs for society and in terms of public guarantees of protection for the citizen. The intervention of the public authority is manifested through the enactment of binding rules governing production practices and, more generally, the organization of the entire supply chain with the aim of achieving an adequate level of food quality and safety. The objective that the authority wants to achieve with this type of initiative is to guarantee the reaching of a minimum quality level, through the creation of Minimum Quality Standards[25] or providing correct information to the final consumer.[26]

The main mandatory public initiatives are listed here below.

#### 5.1.1.1 Regulation (CE) No. 396/2005 of 23 February 2005: Plant Protection Products Maximum Residue Level

The Regulation (CE) No. 396/2005 of the European Parliament and of the Council of 23 February 2005,[27] which entered into force 1 September 2008, lays down harmonized Community provisions on maximum residue levels (MRLs). Annexes II and III of this Regulation set out for the individual active substances contained in plant protection products the maximum residue levels set at Community level in/on the different food matrices listed in Annex I to that Regulation.

#### 5.1.1.2 Council Regulation (CEE) No. 315/93 of 8 February 1993: Contaminants in Foods

Council Regulation (CEE) No. 315/93 of 8 February 1993 laying down Community procedures for contaminants in foodstuffs[28] and refers to a number of regulations and circulars setting maximum levels for contaminants in foods.

#### 5.1.1.3 Commission Regulation (EC) No. 1881/2006 of 19 December 2006: Setting Maximum Levels for Certain Contaminants in Foodstuffs

Commission Regulation (EC) No. 1881/2006 of 19 December 2006 sets maximum levels for certain contaminants (nitrate, mycotoxins, metals, 3-monochloro-1,2-propanediol, polycyclic aromatic hydrocarbons) in foodstuffs in order to protect public health by maintaining the levels of contaminants at toxicologically acceptable levels.[29] Over the years, the Regulation has been repeatedly and profoundly amended through the issuing of specific regulations (in Annex I of this chapter all the valid EU regulation updated to July 2017) because it appears necessary to amend the maximum levels for certain contaminants in order to take account of *Codex Alimentarius* developments.

#### 5.1.1.4 Regulation (EU) No. 1169/2011 of the European Parliament and of the Council of 25 October 2011: Provision Food Information to Consumers

Regulation (EU) No. 1169/2011 on the provision of food information to consumers entered into application on 13 December 2014. The obligation to provide nutrition information is active since 13 December 2016. Mandatory labeling includes the list of ingredients; net quantity; date of minimum durability; special conditions for keeping or use; and the name of the manufacturer, packager, or a vendor established in the Community.

It is noteworthy, though, that provenance must only be indicated when the omission of such information might mislead the consumer; in most cases, it is not required to indicate the origin of products.[30]

### 5.1.2 Voluntary EU Public Initiatives

Public authorities also defined voluntary standards. In these cases, the aim of public intervention is to protect the final consumer as regards the correspondence of the product with respect to certain requirements on which companies base their qualitative differentiation and which therefore aim to ensure the transparency of the prescribed requirements in such a way as to ensure that the quality level is maintained in the long-term.[25]

Consumers are no longer only interested in "organoleptic" attributes, but are increasingly looking for attributes defined earlier as "credence," such as social, ethical, and related to the production techniques of agro-food products, their provenance, the correct remuneration of agricultural producers, respect for the environment, and the correct management of natural resources.

Regulation (EU) No. 1151/2012 of the European Parliament and of the Council of 21 November 2012 on agricultural product and food quality schemes is the new reference text of the European Union framework on food Quality Systems (QS).[24] It repealed the previous legislation on designations of origin, geographical indications, and TSG. The implementing regulations of Commission Regulation (EU) No. 1151/2012 are Commission Delegated Regulation (EU) No. 664/2014 of 18 December 2013 and Commission Implementing Regulation (EU) No. 668/2014 of 13 June 2014.

The QS concerned by this legislation are those which identify, at European level, the names of products with specific qualities linked to a geographical area (PDO and PGI), or the names of products obtained by traditional methods or raw materials (TSG—which are not geographical indications). This identification is carried out following the completion of the procedures and in compliance with the conditions laid down in these regulations, through registration of the name. The registration of names is intended to protect their use: only products that comply with the product specification may be marketed under the registered name. Conversely, all producers who comply with the specification may use the registered name. The virtuous consequences of such a system are the fair gain for producers (they are protected against unfair competition from those who could use the registered name to sell a product of lower quality and lower commercial value) and clear information for consumers (they are given the opportunity to know what they buy because they are assured that the product purchased is obtained in accordance with certain rules contained in the product specification).

The three types of QS mentioned above are briefly described here below.

*Protected Designation of Origin (PDO)*—identifies products that are produced, processed, and prepared in a specific geographical area, using the recognized know-how of local producers and ingredients from the region concerned. These are products whose characteristics are linked to their geographical origin. They must adhere to a precise set of specifications and may bear the PDO logo.

*Protected Geographical Indication (PGI)*—identifies products whose quality or reputation is linked to the place or region where it is produced, processed, or prepared, although the ingredients used need not necessarily come from that geographical area. All PGI products must also adhere to a precise set of specifications and may bear the logo.

*Traditional Speciality Guaranteed (TSG)*—identifies products of a traditional character, either in the composition or means of production, without a specific link to a particular geographical area.

The intensity of the link with the geographical location varies according to whether PDO, PGI, or TSG. For PDO products, an essential or exclusive link between the quality or characteristics of the product and the geographical area of reference is required. All stages of production take place in the area and all raw materials come from the area. There is a direct and exhaustive identification between the area and product. For PGI, the link with the territory can be modulated in a less stringent way. It is only required that a given quality or reputation or other characteristics of the product are attributable to its geographical origin. Only one stage of production must necessarily take place in the geographical area and the raw materials may also come from outside the area. As a rule, the Commission accepts PGI's specifications which provide for the use of raw materials exclusively from the geographical area. However, if there is no obligation to use the raw materials originating in the area. The origin of the raw materials must be left free in accordance with the principle of free movement of goods, unless a restriction on the origin (outside the area) of the raw materials can be justified by the need to maintain the link between the product and the territory. Geographical indications (PDO and PGI) can therefore be defined as particular intellectual property rights, since they are not owned by a person or body but by a territory. The heart of the regulation is in the protection of the name. It is the name that is protected and not the product. The product whose name is protected may be imitated. Therefore, it is not protected as a product. But, if imitated, it cannot be marketed under the protected name. Moreover, the rule also applies to noncomparable products where the use of the registered name is intended to exploit the reputation of the protected name. The protection is extended to the use of PDO and PGI as ingredients. The second rule of the protection regime is to protect the name against any misuse, imitation, or evocation, even if the true origin of the goods or services is indicated or if the protected name is a translation or accompanied by expressions such as «style», «type», «method», «in the manner», «imitation», or similar. The third rule provides for residual protection against any other false or misleading indication as to the source, origin, nature, or essential quality of the product used on the packaging, in advertising material or other documents. This rule is intended to affect cases where the label or the product packaging contains particulars signs or any other visual device intended to manipulate the consumer's perception about the origin, nature, or quality of the product despite the fact that the name under which the product is sold complies with the first and second protection rules.

Despite efforts at Community level to protect local production, EU regulations are very often not enough as there is still a flourishing trend and the uncontrolled marketing of "similar" products, with considerable economic damage to individual consortia.[31] The "similar" products that are sold both in Italy and abroad on the shelves alongside PDO products with names and images on the packaging that somehow evoke the PDO productions (e.g., Gran Grattuggiato, GranMix, etc. made up of a mix of grated cheeses and do not contain Grana Padano cheese inside them) mislead less careful consumers. For years, the various consortia have been complaining about the need to protect the most valuable productions with more specific rules that avoid the confusion of the consumer with stricter requirements as regards images on products and positioning on shelves in supermarkets, but for the moment the actions taken have not been successful.

Up to a 5 years ago, the EU classified wine quality into two categories: Quality Wine Produced in a Specific Region (QWPSR) and Table Wine. These classifications were replaced in 2013 with Regulation (EU) No. 1308/2013 which specify that wines with characteristics attributable to a specific region can be registered under the European Union's quality logos "Protected Designation of Origin" and "Protected Geographical Indication." But they still maintain the specific national quality categories which correspond to PDO and PGI.

The most significant for PDO are:

- France: AOC (Appellation d'Origine Contrôlée);
- Italy: DOC (Denominazione di Origine Controllata) and DOCG (Denominazione di Origine Controllata e Garantita);
- Spain: DO (Denominación de Origen) and DOCa (Denominación de Origen Calificada);
- Portugal: IPR (Indicação de Proveniência Regulamentada) and DOC (Denominacão de Origem Controlada);
- Germany: QbA (Qualitätswein bestimmter Anbaugebiete) and "Prädikatswein" (formerly known as "QmP" or Qualitätswein mit Prädikat);
- Austria: Qualitätswein and Prädikatswein, including DAC (Districtus Austriae Controllatus).

The most significant for PGI are:

- France: VDP (Vin de Pays);
- Italy: IGT (Indicazione Geografica Tipica);
- Spain: VT (Vino de la Tierra);
- Portugal: VR (Vinho Regional);
- Germany: Landwein;
- Austria: Landwein.

### 5.1.2.1 Organic Agriculture

Organic agriculture, sometimes called biological or ecological agriculture, combines traditional conservation-minded farming methods with modern farming technologies. It emphasizes rotating crops, managing pests naturally, diversifying crops and livestock, and improving the soil with compost additions and animal and green manures. Organic farmers use modern equipment, improved crop varieties, soil and water conservation practices, and the latest innovations in feeding and handling livestock. Organic farming systems range from strict closed-cycle systems that go beyond organic certification guidelines by limiting external inputs as much as possible to more standard systems that simply follow organic certification guidelines.[32] The QS of the European Commission also include production from organic farming. In Europe, the production of products from organic farming is regulated by Council Regulation (EC) No. 834/2007. In other countries, organic standards are formulated and overseen by the government. The United States, Canada, and Japan have, like the EU, comprehensive organic legislation, and the term "organic" can be used only by certified producers. In other countries, government guidelines may not exist, and certification is handled by nonprofit organizations and private companies. Internationally, negotiations are underway and some agreements are already in place, to harmonize certification between countries and facilitating international trade. The International Federation of Organic Agriculture Movements (IFOAM) is also working on harmonization.

The EU regulation protects the denominations "biological" and "organic," sets production rules and standards, and defines the procedures of control and inspection.

Typical organic farming practices include[33]:

- wide crop rotation as a prerequisite for an efficient use of on-site resources;
- very strict limits on chemical synthetic pesticide and synthetic fertilizer use, livestock antibiotics, food additives and processing aids, and other inputs;
- absolute prohibition of the use of genetically modified organisms;
- taking advantage of on-site resources, such as livestock manure for fertilizer or feed produced on the farm;
- choosing plant and animal species that are resistant to disease and adapted to local conditions;
- raising livestock in free-range, open-air systems and providing them with organic feed;
- using animal husbandry practices appropriate to different livestock species.

### 5.1.2.2 Quality Terms

In addition to these quality regimes at EU level, there are other optional schemes recognized by the European Commission within the EU Regulation No. 1151/2012 and the implementing and delegated regulations, defined as "Quality terms."

**5.1.2.2.1 Mountain Product** In order to use this term, the products' raw materials and the animal feed used must come essentially from mountain areas, while for processed products, production should generally take place in such areas.

This term was created to enhance the value of mountain products grown or produced in less-favored areas and it is particularly suitable for products of animal origin and honey.

**5.1.2.2.2 Product of EU's Outermost Regions** Outermost regions (French Overseas Departments—Guadeloupe, French Guiana, Réunion and Martinique—and the Azores, Madeira, and the Canary Islands) face difficulties relative to regions in mainland Europe from their remoteness and insularity, including difficult geographical and meteorological conditions. With a view to ensuring greater awareness and consumption of quality agricultural products, whether natural or processed, which are specific to these outermost regions, a logo was introduced in 2006. The regulation sets out specific measures in the agricultural sector to remedy the difficulties caused by the specific situation facing the Union's outermost regions (Fig. 2).

### 5.1.3 Voluntary National Initiatives

The QS recognized at community level, so as reported before, cover designations of origin (PDO and PGI), organic production.

At EU level, different types of QS have been set up to recognize and identify quality production through European/national labels. These labels must be audited and certified by specially authorized independent third parties.

Member States have the possibility to recognize other QS at national level, provided that they ensure quality characteristics that add value to the final product, such as special production techniques, quality criteria that are well above the commercial minimums for the final product, covering issues such as animal welfare, public health, or environmental protection. These QS must be open to all agricultural producers, and must provide production regulations that are binding, transparent to stakeholders and capable of ensuring complete traceability in the production chain.

#### 5.1.3.1 Label Rouge (France)

The Label Rouge is a French national sign of quality assurance as defined by Law No. 2006–11 of the 5 January 2006, which refers to products which by their terms of production or manufacture have a higher level of quality compared to other similar products usually marketed.

# Versions of the graphic symbol

**Fig. 2** Outermost regions logos. Taken from https://ec.europa.eu/agriculture/quality/optional-voluntary-certification_en.

Quality, in this case, refers to all the properties and characteristics of a product that give it its ability to satisfy explicit or implicit needs.

In addition to the sensory characteristics and their perception, of the Label Rouge product by the consumer, the superior quality is based on:

- production conditions, which differ from the conditions of production of usually marketed similar products,
- product image in terms of its conditions of production,
- elements of the presentation or service.

Products which may benefit from a Label Rouge are food and nonfood items, as well as nonprocessed agricultural products (i.e., flowers).

The Label Rouge is open to all products, regardless of their geographical origin (including outside the European Union).

At all stages of its production and its development, the Label Rouge product must meet the requirements defined in the specifications, validated by the Institut national de l'origine et de la qualité (INAO) and approved by a ministerial order published in the Official Journal of the French Republic.

The monitoring of compliance with these requirements and product traceability is ensured by an independent certification body on the basis of a monitoring plan approved by the INAO.

Monitoring of maintenance over time of the high food quality is ensured by performing regular sensory analysis and organoleptic tests that compare the Label Rouge product with the current product.

A commodity or Label Rouge product may benefit simultaneously from a PGI or a TSG, but not an original PDO.[34]

### 5.1.3.2 SQNPI: Sistema di Qualità Nazionale per la Produzione Integrata (Italy)

The Italian Government Law No. 4 of the 3 February 2011, "Disposizioni in materia di etichettatura e di qualità dei prodotti alimentari" in art. 2, subsections 3–9 establishes the National Integrated Production Quality System (hereinafter referred to as SQNPI). The SQNPI quality system is active since January 2016 and is applicable to all crop production.

The SQNPI provides for the adoption of the regional integrated production specifications (approved by the Italian Ministry of Agricultural and Forestry Policy). Integrated production is defined as a system of agro-food production that uses all means of production and defends agricultural production from adversity, aimed at minimizing the use of synthetic chemicals and rationalizing fertilization in compliance with ecological, economical, and toxicological principles. In practice, it is a system of traceability of production to demonstrate that certified products come from farms which apply the above regional specifications. Companies can join SQNPI either individually or in an associated form (e.g., producer consortia, cooperatives, or associations).

The integrated production according to the SQNPI allows on one hand to comply with the legal obligations regarding integrated pest management (National Action Plan) and on the other hand to respond to the pressing market demands of the national and international large-scale retail trade, which is increasingly aware about cultivation methods.

SQNPI, it is a quality system recognized in accordance with Community rules, allows both individual and associated farms to have access to public financing measures under the rural development plans.

### 5.1.3.3 Haute Valeur Environnementale HVE Certification (France)

The haute valeur environnementale (High Environmental Value—HEV) certification comes from the Grenelle Environment law. This environmental certification, which appeared in France in February 2012, is a voluntary approach accessible to all the sectors concerned by four themes: biodiversity, fertilization management, phytosanitary strategy, and water resource management.

The HVE certification is established according to a progressive certification logic taking into account the totality of farms. Environmental certification is controlled by third-party bodies that are both independent and approved by the French Ministry of Agriculture. This type of certification corresponds to the characterization of farms with high environmental performance. The HVE certification cycle is 3 years. Those who have received certification benefit from less pressure from Common Agriculture Policy (CAP) controls and a financial incentive.

This program is similar to the Italian SQNPI and has the same objective. Add to the European regulation of Integrated Pest Management some stringent compliance about environmental sustainability of high-quality farm production.

### 5.1.3.4 VIVA Sustainability and Culture (Italy)

This certification has not already been recognized has quality system following the European rules, because focused only on the wine industry in Italy, but has important innovative aspect.

The VIVA project promoted by the Italian Ministry for the Environment Land and Sea (IMELS) since 2011 aims to improve the sustainability performance of the wine sector, through the analysis of four indicators: Air, Water, Territory, and Vineyard. The scheme uses for the Air indicator the carbon footprint analysis with an LCA approach, while for the Water indicator an innovative methodology that integrates the Water Footprint Network method with an integrated method for gray water.[35] In addition to these first two indicators, which concern environmental sustainability, the program presents two indicators specific for the wine sector: Territory (social, cultural, and economic sustainability) and Vineyard (environmental and social sustainability understood as health of consumers and workers in the agricultural sector).[36]

The program imposes as a condition of access compliance with a series of mandatory minimum criteria linked to the wine sector, both as regards social aspects (e.g., transparency regarding seasonal employment contracts, training of workers, safety in the field, and cellar) and aspects of protection and financing of territorial relaunch activities (the limits are integrated in the Territory indicator).

The certification has a duration of 2 years and at each renewal the company must commit itself to improve its environmental performance through the definition of energy efficiency strategies and environmental impact reduction in the short-, medium-, and long-term in order to undertake a path of continuous improvement.

An interministerial agreement has recently been signed between the Italian Ministry for the Environment and the Italian Ministry for Agricultural Policy, Food and Forestry to coordinate the efforts of the VIVA and SQNPI projects.[37]

### 5.1.4 Voluntary European Initiatives
#### 5.1.4.1 The PEF of the European Commission
The European Commission has launched (as mentioned in the introduction), on an experimental basis, a European program called Single Market for Green Products Initiative, to develop a shared European methodology on the EF of products called PEF, in order to contrast the proliferation of environmental impact assessment methodologies.[12] The testing of the methodology, defined by the Commission Recommendation 2013/179/EU, started in 2013 and will end in early 2018 allowing companies to calculate and communicate the EF of products in a fair and comparable way. The experimentation process involves 25 product categories, 11 of which are linked to the agro-food chain (wine, animal feed, pasta, bottled water, olive oil, red meat, saltwater fish, animal feed for the livestock chain, beer, coffee, and dairy sector). Unfortunately, the number of Italian, Spanish, Portuguese, and Greek companies that participated in this experimental phase was very small. In those few sectors in which they were involved, there was a strong commitment to enhancing the value of local and quality products.

#### 5.1.4.2 EU Ecolabel: Label of Ecological Quality
With Regulation (EC) No. 66/2010, the EU has established a voluntary participatory ecological labeling scheme to promote and make recognizable

products with a reduced environmental impact to consumers throughout their life cycle, to provide consumers with accurate, nondeceptive, and scientifically based information on the environmental impact of products.

Until now, this trademark has expressly excluded its referability to agricultural products and foodstuffs, but Ecolabel is reported here because in the current version of the regulation (that replace the Regulation No. 880/92) there is a reference to the possible extension to food and feed covered by organic farming.

## 5.2 Classification of Private Initiatives

Public regulations often limit themselves to prescribing "result limits" by defining basic parameters or prescribing the mandatory nature of a control system, and are accompanied by private standards of quality and food safety. Private initiatives are mostly undertaken by large organized retail outlets, or by large processing industries which, through the adoption of these standards, implement a system of control over the supply chain, minimizing the possibility of exposure to image risk.

Private initiatives are therefore "more stringent" than public regulation, as they are able to be more "specific," focusing on the critical aspects of the sector analyzed, also intervening in the single sector, defining conditions of effectiveness, specific requirements, and surveillance measures, acting directly on the process.

Increasingly, they also include social and environmental requirements on their suppliers.

Private quality assurance initiatives can be classified on the basis of their intention to signal quality commitment directly to the consumer or not.

### 5.2.1 Standard With a B2B Perspective
#### 5.2.1.1 The GlobalG.A.P. Standard
The GLOBALG.A.P. certificate, also known as the Integrated Farm Assurance standard (IFA), covers good agricultural practices for agriculture, aquaculture, livestock, and horticulture production. It also covers additional aspects of the food production and supply chain such as chain of custody and compound feed manufacturing.

The IFA standard was revised through an extensive stakeholder involvement and consultation process, and IFA standard V5 was published in July 2015 with 1-year conversion period. This means that the IFA standard V5 became obligatory in 2016.

The GLOBALG.A.P. IFA standard V5 is built on a system of modules that enables producers to get certified for several subscopes in one audit. It consists of:

- general regulations: these map out the criteria for successful Control Points and Compliance Criteria (CPCC) implementation, as well as set guidelines for the verification and the regulation of the standard;
- Control Points and Compliance Criteria (CPCC): these clearly define the requirements for achieving the quality standard required by GLOBALG.A.P.

The CPCC are also modular-based consisting of:

- the all farm base module: this is the foundation of all standards and consists of all the requirements that all producers must first comply with to gain certification;
- the scope module: this defines clear criteria based on the different food production sectors. GLOBALG.A.P. covers three scopes: crops, livestock, and aquaculture;
- the subscope module: these CPCC cover all the requirements for a particular product or different aspect of the food production and supply chain.

To get certified, producers must comply with all the CPCC relevant for their subscope.[38]

### 5.2.1.2 The British Retail Consortium Global Standard

The British Retail Consortium (BRC) global standards are a private, for profit, membership-based association. It represents the whole range of retailers for the UK retail industry, from the large department stores through to independents. Over the past 13 years, BRC has developed the BRC global standards, a suite of four industry-leading technical standards that specify production, packaging, storage, and distribution requirements to guarantee safe food and consumer products. Originally developed in response to the needs of UK members of the BRC, the standards have gained usage worldwide and are specified by retailers and branded manufacturers in the EU, North America, and elsewhere.

### 5.2.1.3 International Featured Standard: IFS 6.1 Food

The IFS Food Standard is a GFSI (Global Food Safety Initiative) recognized standard for auditing food manufacturers. The focus is on food safety and the quality of processes and products. It concerns food processing companies and companies that pack loose food products.

IFS Food applies when products are "processed" or when there is a hazard for product contamination during primary packing. The Standard is important for all food manufacturers, especially for those producing private labels, as it contains many requirements related to the compliance with customer specifications.

IFS Food has been developed with full and active involvement of certification bodies, retailers, food industry, and food service companies.[39]

### 5.2.1.4 ISO 22000: Food Safety Management

The International Organization for Standardization (ISO) developed the food safety management system certification: ISO 22000. ISO and its member countries used the quality management system approach and tailored it to apply to food safety, incorporating the widely used and proven Hazard Analysis and Critical Control Points (HACCP) principles and good manufacturing principles. The standard has requirements for food safety management systems processes and procedures, and requires that the organization implement prerequisite programs and HACCP.

Unlike some of the other food safety management systems certification programs (for example, FSSC 22000 and SQF), the ISO 22000 does not have specific requirements for PRerequisite Programs (PRPs), but requires that the organization identifies and implements the appropriate programs. This makes it more flexible, and food organizations of any type can implement and be certified to ISO 22000.

Food processors and manufacturers can use the ISO technical specification ISO/TS 22002-1 to develop their PRPs. It outlines the requirements for PRPs that are applicable to these organizations.[40]

This ISO is now under revision and the updated version will be released at the end of this year.

### 5.2.1.5 Safe Food Quality Institute: SQF Code

The SQF code was redesigned in 2012 for use by all sectors of the food industry from primary production to transport and distribution. It replaced the SQF 2000 code edition 6 and the SQF 1000 code edition 5.

The SQF code is a process and product certification standard. It is a HACCP-based food safety and quality management system that utilizes the National Advisory Committee on Microbiological Criteria for Food (NACMCF) and the *CODEX Alimentarius* Commission HACCP principles and guidelines, and is intended to support industry or company branded product and to offer benefits to suppliers and their customers. Products produced and manufactured under the SQF code certification retain a high degree of acceptance in global markets.

First developed in Australia in 1994, the SQF program has been owned and managed by the Food Marketing Institute (FMI) since 2003, and was recognized (at level 2) in 2004 by the Global Food Safety Initiative (GFSI) as a standard that meets its benchmark requirements. The SQF code level 3 exceeds the requirements of the GFSI benchmark documents.

The main feature of the SQF code is its emphasis on the systematic application of HACCP for control of food quality hazards as well as food safety. The implementation of an SQF management system addresses a buyer's food safety and quality requirements and provides the solution for businesses supplying local and global food markets.

Certification of an SQF system by a certification body licensed by the Safe Quality Food Institute (SQFI) is not a statement that the certification body guarantees the safety of a supplier's food or service, or meets all food safety regulations at all times. However, it is an assurance that the supplier's food safety plans have been implemented in accordance with the HACCP method and applicable regulatory requirements and that they have been verified and determined effective to manage food safety. It is also a statement of the supplier's commitment to[41]

1. produce safe, quality food;
2. comply with the requirements of the SQF code; and
3. comply with applicable food legislation.

### 5.2.2 Standard With a B2C Perspective

#### 5.2.2.1 ISO 14020 Family of Environmental Labeling

##### 5.2.2.1.1 Environmental Labels and Declarations: Type I: ISO 14024:2001 This International Standard establishes the principles and procedures for developing Type I environmental labeling programs, including the selection of product categories, product environmental criteria, and product function characteristics, and for assessing and demonstrating compliance. The standard also establishes the certification procedures for awarding the label. Type I environmental labeling refers to the multicriteria, life cycle seals of approval, commonly known as "ecolabeling." ISO 14024 provides the requirements for operating an ecolabeling scheme, like the Nordic Swan or the Japanese Eco-Mark.

The principles of this standard include the following stipulations:
- environmental labeling programs should be voluntary;
- compliance with environmental and other relevant legislation is required;

- the whole product life cycle must be taken into consideration when setting product environmental criteria, e.g., extraction of resources, manufacturing, distribution, use, and disposal relating to relevant crossmedia environmental indicators. Any departure from this comprehensive approach or selective use of restricted environmental issues has to be justified;
- product environmental criteria need to be established to differentiate environmentally preferable products from others in the product category when these differences are significant;

### 5.2.2.1.2 Environmental Labels and Declarations: Type II: ISO 14021:2016 ISO 14021:2016 is the International Standard that deals with so-called self-declared claims. It states that the overall goal of environmental labels and declarations is, through the communication of verifiable, accurate information that is not misleading, to encourage the demand for, and supply of, products which cause less stress on the environment, thereby stimulating the potential for market-driven, continual environmental improvement.

ISO 14021 addresses the issue that if a claim is made, how can it be made in a way which is meaningful and useful to a consumer. The objectives of ISO 14021 are stated to be the harmonization of the use of self-declared environmental claims with the following anticipated benefits:

- accurate and verifiable environmental claims that are not misleading;
- increased potential for market forces to stimulate environmental improvements;
- prevention or minimization of unwarranted claims;
- reduction in marketplace confusion;
- facilitation of international trade;
- increased opportunity for consumers to make more informed choices.

There are three key elements to ISO 14021 concerning requirements for claims. These give the basic rules for the making of environmental claims.

- use of symbols: this deals with the fact that many claims for products are made not just with the use of text but also by the use of pictures, symbols, or logos;
- evaluation and claim verification requirements: essentially this requires that claims must be verified before they are made, and that this information must be available on request to any person;
- specific requirements for selected claims: this recognizes that some claims are used more frequently than others (e.g., recyclable or biodegradable) and provide for specific requirements in the use of such claims.

The basic requirements for all claims are that they shall be:

- accurate and not misleading;
- substantiated and verified;
- unlikely to result in misinterpretation.

Italian government with the Legislative Decree No. 50 of 18 April 2016 has enhanced the value of Type II declarations in the minimum environmental criteria (CAM in Italy) for green public procurement.

If, on the one hand, this decree opens up a new horizon for the use and diffusion of environmental declarations, on the other hand, it opens up a problem linked to the impossibility and nonexistence, at present, of a certification and accreditation system according to ISO 14021, which currently regulates self-declared environmental declarations. In the following years, the Italian Accreditation Board—ACCREDIA will have to deal with this legislative evolution by defining rules or guidelines, at least national ones, for the certification or verification of Type II environmental statements.

### 5.2.2.1.3 Environmental Labels and Declarations: Type III: ISO 14025:2010 ISO 14025 establishes principles and specifies procedures for issuing quantified environmental information about products, based on life cycle data referred to as environmental declarations. A Type III environmental declaration can be described as: quantified environmental data for a product with preset categories of parameters based on the ISO 14040 series of standards, but not excluding additional environmental information.

Type III environmental declarations present the environmental performance of a product to enable objective comparisons between products fulfilling the same function.

Such declarations:

- are based on independently verified Life Cycle Assessment (LCA) data, Life Cycle Inventory analysis (LCI) data, converted LCI data to reflect the Life Cycle Impact Assessment (LCIA) of a product or information modules in accordance with the ISO 14040 series of standards and, where relevant, additional environmental information;
- are developed using predetermined parameters;
- are subject to the administration of a program operator, such as a company or a group of companies, industrial sector or trade association, public authorities or agencies, or an independent scientific body or other organization.

The output is known as the declaration which has the objective to communicate environmental performance within clearly defined and classified product categories and service types. The system approach covers separate products and services, as well as complete or partial assortments of products and services. A Type III environmental declaration is designed to meet various information needs within the supply chain and for end products in both the private and public sectors, as well as for more general purposes in information activities and marketing.

### 5.2.2.1.4 The International Environmental Product Declaration System
The International Environmental Product Declaration (EPD) System is a program that provides EPDs in accordance with ISO 14025 and ISO/TS 14067 standards; it helps and supports companies in communicating the environmental performance of their products in a credible and comprehensible way.[42] The program, with technical secretariat located in Sweden, has an international focus. The EPD of Italian agro-food products are about 125 in the International EPD System database.[43]

## 6. FOCUS ON GEOGRAPHICAL INDICATIONS

At the end of 2017 there were a total of 3005 PGI, TSG, and PDO products in EU and non-EU countries, of which 1419 food products (47%) and 1586 wine products (53%). Food products in the EU countries are divided into 623 PDO (45%), 714 PGI (51%), and 56 TSG (4%) with respect to the type of certification, while wines are divided into 1148 PDO (72%) and 438 PGI (28%). A total of 46 new products were registered in 2017, 39 in the food and 7 in the wine segment. For the food sector, 39 new names have been entered in the register of PDO, PGI, and TSG branded products, of which 12 are PDO, 25 PGI, and 2 TSG. Among these new products, three are related to non-EU countries (Indonesia, Turkey, and Norway). The new products registered mainly belong to the categories meat products (+8), fruit, vegetables, and cereals (+7), fresh meat (+6), other products reported in the Annex I (+5), followed by cheese, fish, and molluscs (+4). With the exception of the three products from outside the EU, since the beginning of the year there have been 36 registered food denominations from European countries, and in comparison to the total of December 2016 (1357) the food sector registered an increase of +2.7% in 2017. In addition, in 2017, the wine segment also had seven new products (it was since 2013, with the registration of Croatian wines, that the sector showed no change in terms of new names): these were seven PDO wines registered in France (3), Germany (1), the United Kingdom (1), Slovakia (1), Belgium and the Netherlands (1) (Table 1).

**Table 1** List of EU Tradition and Quality Products[44,45]

| Country | Food | | | | Wine | | |
|---|---|---|---|---|---|---|---|
| | POD | PGI | STG | Total | PDO | PGI | Total |
| Italy | 167 | 126 | 2 | 295 | 474 | 129 | 603 |
| France | 103 | 141 | 1 | 245 | 379 | 75 | 454 |
| Spain | 102 | 89 | 4 | 195 | 100 | 45 | 145 |
| Greece | 76 | 30 | 0 | 106 | 33 | 116 | 149 |
| Portugal | 64 | 73 | 1 | 138 | 46 | 10 | 56 |
| Germany | 12 | 77 | 0 | 89 | 14 | 26 | 40 |
| Hungary | 6 | 8 | 1 | 15 | 54 | 8 | 62 |
| The United Kingdom | 25 | 41 | 4 | 70 | 3 | 2 | 5 |
| Bulgaria | 0 | 2 | 5 | 7 | 52 | 2 | 54 |
| Romania | 1 | 3 | 0 | 4 | 38 | 13 | 51 |
| Czech Republic | 6 | 23 | 4 | 33 | 12 | 2 | 14 |
| Austria | 10 | 6 | 1 | 17 | 26 | 3 | 29 |
| Slovenia | 8 | 13 | 3 | 24 | 14 | 3 | 17 |
| Slovakia | 2 | 10 | 7 | 19 | 18 | 3 | 21 |
| Poland | 8 | 22 | 9 | 39 | 0 | 0 | 0 |
| Croatia | 10 | 9 | 0 | 19 | 16 | 0 | 16 |
| Belgium | 3 | 11 | 5 | 19 | 8 | 2 | 10 |
| The Netherlands | 6 | 5 | 4 | 15 | 1 | 12 | 13 |
| Cyprus | 1 | 4 | 0 | 5 | 7 | 4 | 11 |
| Denmark | 0 | 7 | 0 | 7 | 0 | 4 | 4 |
| Finland | 5 | 2 | 3 | 10 | 0 | 0 | 0 |
| Sweden | 3 | 3 | 2 | 8 | 0 | 0 | 0 |
| Lithuania | 1 | 4 | 2 | 7 | 0 | 0 | 0 |
| Ireland | 3 | 4 | 0 | 7 | 0 | 0 | 0 |
| Latvia | 1 | 1 | 3 | 5 | 0 | 0 | 0 |
| Luxembourg | 2 | 2 | 0 | 4 | 1 | 0 | 1 |
| Malta | 0 | 0 | 0 | 0 | 3 | 1 | 4 |
| Estonia | 0 | 0 | 0 | 0 | 0 | 0 | 0 |

## 7. MULTIPLICATION OF ENVIRONMENTAL AND SUSTAINABILITY LABELS

The consumer interest in sustainability performance of products is also proved by the constant growth of sustainability certification schemes. The multiplication of sustainability schemes has been observed since the 1990s,[11,46,47] Gruère observed a multiplication of the Organization for Economic Cooperation and Development (OECD) countries total number of schemes by five (from nearly 100 in 1990 to more than 500 schemes in 2013). The food and agriculture sector, with textile, forestry, and building have a dominant share of schemes.

This proliferation easily can lead to consumer confusion and frustration. Fischer and Lyon[48] explain that the label multiplication can impair environmental performance in general, even when consumers have access to perfect information, and that consumer confusion makes matters worse.

The proliferation of methodologies is hampering the functioning of the green products market. To be able to fulfill high market expectations and to reduce the number of misleading or false sustainability claims, increasing the number of correct ones is important to define common international standards. Groups of products usually differ in their inherent sustainability performance so beside internationally agreed standards also common and harmonized calculation rules have to be established within the same product group, to ensure that similar procedures are used. Unfortunately, the definition of shared and common methods can be not enough and is proved that small amounts of uncertainty about the quality of labels can create consumer confusion that reduces or eliminates the value to firms adopting them. Label multiplication aggravates the effect of uncertainty, and as the number of labels becomes large, labeling becomes completely uninformative.[49] This study suggests that if one label becomes "focal" (i.e., whether this is a standard expected by consumers), this can alleviate the problem of multiplication. Failure to adopt the focal label is taken by consumers as a significant failing and has a greater impact on sales than if the label was just one among many options. In fact, in a world with label uncertainty, a focal label plus multiple uncertain labels actually performs better than a single label, because firms that do not meet the standard of the focal label can still convey some information to consumers by meeting one of the others. This suggests that a government-backed label may be able to bring some order to a chaotic labeling market.

## 8. EUROPEAN CONSUMER BEHAVIOR

Sales of PDO and PGI products and products from organic farming are growing steadily in recent years, mainly due to the fact that local and organic products may be categorized as "sustainable produced food" since they reflect two different components of sustainability: a social component related to the integration of the support of the agro-food sector with the priorities and needs of citizens and an environmental component relating to sustainable use and management of the natural resources.[50]

In particular, Organic Agriculture has had and is still having an extraordinary success. Consumers prefer organic production despite the higher costs. The key to this success is linked to the ability of these products to evoke in consumers a multiplicity of "emotions." Organic products are perceived as having both attributes of "public benefit" (e.g., reduced environmental impact, animal welfare, biodiversity, and protection of small and local producers) but also by attributes of "private benefit" because they are perceived as products that guarantee a healthier and "taster" diet.[51,52] Regardless of the type of benefit that the consumer is looking for, organic products are perceived as an optimal mix of quality, health, and environmental protection.

These trends are reflected also by the European Commission data related to the concerns of European citizen. In 2009, according to the European Commission's barometer,[53] nearly half of Europeans (47%) ranked climate change as the second most important issue, behind poverty and ahead of the global recession. The concept of "sustainable development" continues to find its way into people's minds, and not only is awareness is growing, citizens also feel them part of the process, are increasingly sensitive to certain environmental problems, and therefore take them into account in their daily choices.[54]

Various studies carried out on European consumers prove that: French citizens are widely aware of certain environmental issues,[55] 93% of them agree that individual efforts can have a significant impact on environmental protection, and 78% agree that they could do more or better in their daily choices to ensure the compliance with a more sustainable development. Those most committed to the environmental cause are also the most convinced that they could still change their daily practices. But even among those who are more indifferent to environmental issues, almost 7 in 10 feel that they could do better or more on a daily basis to ensure sustainable development.[54]

In Italy, consumers willing to pay a premium price for sustainable brands are 52%, up sharply from 44% in 2013, and 45% in 2014.[56]

In Germany, TNS Emnid conducted a population-representative survey on behalf of the federation of German consumer organizations (vzbv) in January 2016. The key results of the study are that two-thirds of consumers are (always) willing to buy sustainable products when shopping groceries, 63% claim to not find sufficient information on the sustainable footprint of articles and 36% have difficulties in recognizing sustainable products yet. Further, 55% criticize high prices and 44% the poor availability of sustainable food products.[57]

In Europe, consumers willing to pay a premium price for sustainable brands in 2015 are 51% (40% in 2014 and 37% in 2013), and at global level, the figure rises to 66%, growing faster by 11 percentage points compared to 2014 and by 16 points from 2013.[56] At global level, companies committed to environmental and social sustainability registered a 4% growth in 2015 turnover, unlike those not engaged, whose turnover increased by <1%.[56]

## 9. CONCLUSIONS AND FUTURE DEVELOPMENTS

EU agro-food products, both as agricultural raw materials and final products, enjoy an undisputed and envied reputation of the highest quality on the world market, especially for Italian products. The good reputation of European production at global level is also due to the respect of minimum requirements set at technically high levels (protection of health and the environment and animal welfare) compared to production by international competitors, which help to give European agro-food production an added value. However, compliance with the minimum requirements is no longer sufficient, especially for European consumers, and the interest in products with "credence" characteristics is constantly growing. The market is increasingly subject to consumer demand for detailed information on the production criteria and origin of products, especially in relation to specific themes or issues of high media interest (e.g., environmental statements such as: Programme for the Endorsement of Forest Certification schemes—PEFC, Forest Stewardship Council—FSC, Dolphin Safe the international dolphin safe scheme for the protection of dolphins in tuna fishing, products with sustainable palm oil, and products of animal origin without antibiotics). This leads companies to differentiate themselves from their national and international competitors, creating or embracing programs, specific for product categories, which provide direct information to the final consumer. But consumers, being flooded with environmental information and a multitude of programs, struggle to choose or understand which program is the most reliable and best responds to their needs, causing a distrust of certifications and environmental declarations.

As mentioned in the previous paragraph, the only quality programs that are free of this confusion are PDO and PGI products and organic farming products, precisely because of their ability to evoke a multiplicity of positive "emotions" in the final consumer. Unfortunately, however, the sustainability performance of these products is often not assessed. The production specifications and the regulations for organic farming do not provide for a company commitment to more sustainable production, while the consumer is also aware of social and environmental quality aspects for these products. This means that the companies producing EU tradition and quality products and organic products are not stimulated to make their production system more sustainable, but aim at maximizing production in compliance with the limits of their specifications.

It is therefore clear that the European Commission has not yet been able to find a successful synthesis between the different quality cores. It has only recently started thinking about regulations on voluntary aspects of integration between quality and sustainability (see single market for green products: PEF and OEF), but at present the existing programs are not consistent and in some cases overlap with some objectives (i.e., ecolabel and PEF).

Individual member states, already some years ago, have started to develop national programs in this direction (i.e., HVE, SQNPI, and VIVA). A concrete example that is trying to integrate the quality of a product such as wine, strongly linked in Italy to PDO and PGI productions, with sustainability requirements is the VIVA program for sustainability in Italian wine growing. VIVA is one of the first attempts in Europe to focus on a specific sector, identifying the most sensitive issues at national level and of interest to the final consumer, placing them as requirements for access to the scheme through the territory indicator. The program is therefore an optimal combination of environmental performance analysis and implementation of mitigation strategies and actions, with minimum access requirements evolving over the years. The project is not limited to identifying wines with better environmental performance, but promotes and directs the Italian wine sector to a path of innovation and improvement.

Italy now, through the establishment of the MGI scheme, intends to develop the VIVA experience in a national scheme that integrates the PEF methodology of the European Commission with additional information and/or will establish minimum criteria for program access to enhance sustainability aspects related to traceability, landscape quality, and social sustainability. The definition of specific Product Environmental Footprint Category Rules (PEFCR) adopted by the PEF program (and by the MGI) is well integrated with what has been discussed, as the structure makes it possible to address and understand the specific problems of each sector analyzed.

In the definition of PEFCR at European level, some stakeholders have come together under a single technical secretariat in order to define specific rules for the assessment of environmental impacts in accordance with the PEF methodology. In the same way, therefore, it will be possible to define critical aspects of social and landscape sustainability that the product category will have to respect and any additional information, which can be integrated in the sustainability declaration attached to the study. The development of "broadened" participatory processes involving companies, consumer associations, and stakeholders in the supply chain gives greater legitimacy to the standard developed, while at the same time raise the bar of voluntary legislation.

The European production system fits well with this type of scheme and must aim to develop it. To confirm this, it is enough to look at the data regarding farms, which in the last decade have focused extensively on multifunctionality. Italy, with an average of 4.6 billion €, ranks first in the EU for support service activities, followed by France (4.2 billion €) and Germany (2.1 billion €). In the field of support activities, labor subcontracting, first processing of products, and maintenance of the territory are emerging. These aspects are also interesting in order to enhance secondary activities such as the production of renewable energy and the spread of agritourism. Italy is in first place with 27.5% of European production, followed by France (14.2%) and the United Kingdom (9.9%).[58]

The last point that has not yet been dealt with, but which is certainly of interest to companies in the European agricultural and agro-food sector, concerns the predominantly economic limits that such complex schemes may have. Until now, in fact, the main difficulties that these schemes, which integrate more aspects of sustainability, had to deal with were higher costs related to the costs of carrying out studies and certification, as well as having to define optimal strategies of communication to the consumer in order to avoid being compared to scientifically unsound schemes.

Surely the evolution of methodologies is making this type of study more competitive from the point of view of the cost that companies have to bear. PEF for how it is structured, and consequently the MGI, allows consortia or groups of companies to merge with the objective of building tools that simplify the application of LCA studies, significantly reducing the costs associated with analysis and certification and allowing small- and medium-sized enterprises (SMEs) to make these studies affordable.

The European Commission is already funding research in this area, although national and local authorities have a key role to play, as they will need to encourage and support the development and implementation of these sustainability schemes through calls for proposals for agricultural funding and future policies.

It is clear that only the generalized entry on these issues of public institutions as managers and innovators in regulatory matters, with a central role for the European institutions, which have always been progressive in this area, can guarantee a fair mediation between the collective interests and the interests of companies, putting consumer protection at the center.

The fundamental objective of quality is to inform the final consumer about the specific characteristics of the product. Satisfaction with consumer expectations is one of the cornerstones of the ISO definition of quality, which presupposes a conscious and well-informed purchase.

The process of reforming the quality of agricultural and agro-food production cannot and must not stop. It is still necessary for both public institutions and voluntary private sector schemes to achieve four objectives:

1. improve communication between producers, buyers, and consumers on the quality of agricultural and agro-food products;
2. making EU quality policy instruments more coherent;
3. simplify label wording: the European Commission has not yet been able to find a proper synthesis between the different quality cores, investing also in consumer understanding;
4. integrating sustainability schemes more directly with CAP policies.

There is therefore an urgent need to enhance the value of these quality products with more consistent approaches, which include crosscutting provisions with shared levels of quality with civil society and verification instruments and regulations that are suitable for preventing abuse and that support the correct information of the consumer.

## ANNEX I

Valid EU regulation updated to July 2017 setting maximum levels for certain contaminants in foodstuffs:

- Commission Regulation (EU) No. 2017/1237 of 7 July 2017 amending Regulation (EC) No. 1881/2006 as regards the maximum level of hydrocyanic acid in unprocessed whole, ground, ground, mololite, crushed, and ground apricot seeds placed on the market for the final consumer.

- Commission Regulation (EU) No. 2016/239 of 19 February 2016 amending Regulation (EC) No. 1881/2006 as regards maximum levels for tropical alkaloids in certain processed cereal-based foods for infants and young children.
- Commission Regulation (EU) No. 1940/2015 of 28 October 2015 amending Regulation (EC) No. 1881/2006 as regards maximum levels for sclerotia of *Claviceps* spp. in certain unprocessed cereals and monitoring and reporting requirements.
- Commission Regulation (EU) No. 1933/2015 of 27 October 2015 amending Regulation (EC) No. 1881/2006 as regards maximum levels for polycyclic aromatic hydrocarbons in cocoa fiber, banana chips, food supplements, dried aromatic herbs, and dried spices.
- Commission Regulation (EU) No. 1137/2015 of 13 July 2015 amending Regulation (EC) No. 1881/2006 as regards the maximum level of ochratoxin A in spices *Capsicum* spp.
- Commission Regulation (EU) No. 1006/2015 of 25 June 2015 amending Regulation (EC) No. 1881/2006 as regards the maximum levels for inorganic arsenic in foodstuffs.
- Commission Regulation (EU) No. 1005/2015 of 25 June 2015 amending Regulation (EC) No. 1881/2006 as regards the maximum levels for lead in certain foodstuffs.
- Commission Regulation (EU) No. 704/2015 of 30 April 2015 amending Regulation (EC) No. 1881/2006 as regards the maximum level of nondioxin-like PCBs in wild spinarolo (*Squalus acanthias*).
- Commission Regulation (EU) No. 1327/2014 of 12 December 2014 amending Regulation (EC) No. 1881/2006 as regards the maximum levels for polycyclic aromatic hydrocarbons (PAH) in traditionally smoked meat and meat products, as well as in traditionally smoked fish and fishery products.
- Commission Regulation (EU) No. 696/2014 of 24 June 2014 amending Regulation (EC) No. 1881/2006 as regards maximum levels for erucic acid in vegetable oils and fats and in foodstuffs containing vegetable oils and fats.
- Commission Regulation (EU) No. 488/2014 of 12 May 2014 amending Regulation (EC) No. 1881/2006 as regards the maximum levels for cadmium in foodstuffs.

- Commission Regulation (EU) No. 212/2014 of 6 March 2014 amending Regulation (EC) No. 1881/2006 as regards maximum levels for the contaminant citrinin in food supplements based on fermented rice with *Monascus purpureus* red yeast.
- Commission Regulation (EU) No. 1067/2013 of 30 October 2013 amending Regulation (EC) No. 1881/2006 as regards maximum levels for dioxin, dioxin-like PCBs, and nondioxin-like PCBs in the liver of terrestrial animals.
- Commission Regulation (EU) No. 1058/2012 of 12 November 2012 amending Regulation (EC) No. 1881/2006 as regards maximum levels for aflatoxins in dried figs.
- Commission Regulation (EU) No. 594/2012 of 5 July 2012 amending Regulation (EC) No. 1881/2006 as regards maximum levels for ochratoxin A, nondioxin-like PCBs, and melamine in foodstuffs.
- Commission Regulation (EU) No. 1259/2011 of 2 December 2011 amending Regulation (EC) No. 1881/2006 as regards maximum levels for dioxin-like PCBs and nondioxin-like PCBs in foodstuffs.
- Commission Regulation (EU) No. 1258/2011 of 2 December 2011 amending Regulation (EC) No. 1881/2006 as regards maximum permitted levels for nitrates in foodstuffs.
- Commission Regulation (EU) No. 835/2011 of 19 August 2011 amending Regulation (EC) No. 1881/2006 as regards maximum levels for polycyclic aromatic hydrocarbons in foodstuffs.
- Commission Regulation (EU) No. 420/2011 of 29 April 2011 amending Regulation (EC) No. 1881/2006 setting maximum levels for certain contaminants in foodstuffs.
- Commission Regulation (EU) No. 165/2010 of 26 February 2010 amending Regulation (EC) No. 1881/2006 setting maximum levels for certain contaminants in foodstuffs as regards aflatoxins.
- Commission Regulation (EU) No. 105/2010 of 5 February 2010 amending Regulation (EC) No. 1881/2006 setting maximum levels for certain contaminants in the products of the product concerned.

# ANNEX II

| Name | Description | Established in | Managing body | Regulatory Process | Implementation | Geographic Perimeter | Control | Target |
|------|-------------|----------------|---------------|-------------------|----------------|----------------------|---------|--------|
| Regulation (CE) No. 396/2005 | Plant Protection products maximum residue level | 2005 | EU institution | Mandatory | Effective | All products sold in EU | Guaranteed by the national state | B2B |
| Council Regulation (CEE) No. 315/93 | Contaminants in foods | 1993 | EU institution | Mandatory | Effective | All products sold in EU | Guaranteed by the national state | B2B |
| Commission Regulation (EC) No. 1881/2006 | Setting maximum levels for certain contaminants in foodstuffs | 2006 | EU institution | Mandatory | Effective | All products sold in EU | Guaranteed by the national state | B2B |
| Regulation (EU) No. 1169/2011 | Provision food information to consumers | 2011 | EU institution | Mandatory | Effective | All products sold in EU | Guaranteed by the national state | B2C |
| Regulation (EU) No. 1151/2012 | Quality schemes for agricultural products and foodstuffs (PDO, PGI, TSG, mountain product, product of EU's outermost regions) | 2012 | EU institution | Voluntary | Effective | Worldwide | Guaranteed by the national state | B2B B2C |

Continued

| Name | Description | Established in | Managing body | Regulatory Process | Implementation | Geographic Perimeter | Control | Target |
|---|---|---|---|---|---|---|---|---|
| Organic farming | Biological or ecological agriculture that combines traditional conservation-minded farming methods with modern farming technologies | 1924[a] | National public institution and NGO's | Voluntary | Effective | Worldwide | Independent certification | B2B B2C |
| Label Rouge—French Law No. 2006–11 | Products which by their terms of production or manufacture have a higher level of quality compared to other similar products usually marketed | 2006 | French public institution | Voluntary | Effective | Worldwide | Independent Certification | B2C |
| SQNPI—Italian Law No. 4 of the 3 February 2011 | Provides for the adoption of the regional integrated production specifications approved by the Ministry of Agricultural and Forestry Policy | 2011 | Italian public institution | Voluntary | Effective | Italy | Independent Certification | B2C |
| Haute valeur environnementale French certification issued by Grenelle Environnement conference | Approach accessible to all the sectors concerned by four themes: biodiversity, fertilization management, phytosanitary strategy, and water resource management | 2012 | French public institution | Voluntary | Effective | France | Independent Certification | B2C |

| Name | Description | Established in | Managing body | Regulatory Process | Implementation | Geographic Perimeter | Control | Target |
|------|-------------|----------------|---------------|--------------------|----------------|----------------------|---------|--------|
| VIVA Sustainability and Culture | Program to assess the sustainability performance of the wine sector in Italy | 2011 | Italian public institution | Voluntary | Effective | Italy | Independent Certification | B2C |
| PEF-OEF Product and Organization Environmental Footprint Commission Recommendation 2013/179/EU | Environmental footprint impact assessment methodology based on Life Cycle Assessment | 2013 | EU institution | Voluntary | Experimental phase open for a 2 year test from mid-2018 | All products sold in EU for which a PEF-OEF Category Rules has been drawn up | Independent Certification | B2B B2C |
| EU Ecolabel—Regulation (EC) No. 66/2010 | Ecological labeling scheme to promote and make recognizable products with a reduced environmental impact | 1992 | EU institution | Voluntary | Effective | All products sold in EU for which a EU Ecolabel criteria has been drawn up | Independent Certification | B2C |
| The Global GAP standard (IFA Standard V5) | Integrated Farm Assurance standard for agriculture, aquaculture, livestock, and horticulture production | 1997 as Eurepgap | Private | Voluntary | Effective | Worldwide | Independent Certification | B2B |

Continued

| Name | Description | Established in | Managing body | Regulatory Process | Implementation | Geographic Perimeter | Control | Target |
|---|---|---|---|---|---|---|---|---|
| The British Retail Consortium (BRC) global standard | Four industry-leading technical standards that specify production, packaging, storage, and distribution requirements to guarantee safe food and consumer products | 1992 | Private | Voluntary | Effective | Worldwide | Independent Certification | B2B |
| International Featured Standard—IFS 6.1 food | Global Food Safety standard for auditing food manufacturers focussed on food safety and quality of processes and products | 2003 | Private | Voluntary | Effective | Worldwide | Independent Certification | B2B |
| ISO 22000—Food safety management | Food safety management system certification using HACCP | 2005 | Private | Voluntary | Effective | Worldwide | Independent Certification | B2B |
| Safe Food Quality Institute—SQF code | Food safety and quality management system HACCP based | 2009 | Private | Voluntary | Effective | Worldwide | Independent Certification | B2B |
| Environmental labels and declarations—Type I—ISO 14024:2001 | Requirements for operating an ecolabeling scheme | 2001 | Private | Voluntary | Effective | Worldwide | Independent Certification | B2C |

| Name | Description | Established in | Managing body | Regulatory Process | Implementation | Geographic Perimeter | Control | Target |
|---|---|---|---|---|---|---|---|---|
| Environmental labels and declarations—Type II—ISO 14021:2016 | Standard for self-declared environmental claims | 2001 | Private | Voluntary | Effective | Worldwide | Independent Certification | B2C |
| Environmental labels and declarations—Type III—ISO 14025:2010 | Specifies procedures for issuing quantified environmental information about products, based on life cycle assessment | 2006 | Private | Voluntary | Effective | Worldwide | Independent Certification | B2B B2C |

[a]Developed in 1924 with the Biodynamic agriculture of Rudolf Steiner.

# REFERENCES

1. Aprile MC, Caputo V, Nayga RM. Consumers' valuation of food quality labels: the case of the European geographic indication and organic farming labels. *Int J Consum Stud* 2012;**36**:158–65. https://doi.org/10.1111/j.1470-6431.2011.01092.x.
2. European Commission. *Food and drink industry.* WWW Document, https://ec.europa.eu/growth/sectors/food_it; 2017. (accessed 12.18.17).
3. Eurostat. *Farm structure statistics.* WWW Document, http://ec.europa.eu/eurostat/statistics-explained/index.php/Farm_structure_statistics; 2015. (accessed 12.18.17).
4. Infomercatiesteri.it. *Scheda di sistesi settore agroalimentare.* WWW Document, http://www.infomercatiesteri.it/ranking_scheda_sintesi.php?id_settori=1; 2017. (accessed 12.13.14).
5. ISTAT, 2014. Annuario statistico Italiano C13 Agricoltura.
6. Nielsen Global Survey, 2016. What's in our food and mind-ingredients and dining out trends around the world.
7. Prepareed Foods. *Nielsen survey: consumer eating habits.* WWW Document, https://www.preparedfoods.com/articles/118731-nielsen-survey-consumer-eating-habits; 2016.
8. Eurostat. *Agriculture—greenhouse gas emission statistics.* WWW Document, http://ec.europa.eu/eurostat/statistics-explained/index.php/Agriculture_-_greenhouse_gas_emission_statistics; 2015. (accessed 12.13.17).
9. Regioni.it, 2017. Bonn: la Conferenza sul clima. Reg 3267.
10. European Commission, 2011. COMUNICAZIONE DELLA COMMISSIONE AL PARLAMENTO EUROPEO, AL CONSIGLIO, AL COMITATO ECONOMICO E SOCIALE EUROPEO E AL COMITATO DELLE REGIONI—Tabella di marcia verso un'Europa efficiente nell'impiego delle risorse—Com (2011) 571. Bruxelles doi: https://doi.org/10.1017/CBO9781107415324.004.
11. European Commission, 2013a. Impact Assessment Report for the COMMUNICATION FROM THE COMMISSION TO THE EUROPEAN PARLIAMENT AND THE COUNCIL Building the Single Market for Green Products: facilitating better information on the environmental performance of products and organisations.
12. European Commission. *Single Market for Green Product initiative.* WWW Document, http://ec.europa.eu/environment/eussd/smgp/; 2013. (accessed 12.5.17).
13. Council of the European Union. *Council conclusions on eco-innovation: enabling the transition towards a circular economy.* in: PRESS RELEASE vol. 18 December 2017. Brussels, 2017. p. 2–3.
14. Nelson P. Information and consumer behavior. *J Polit Econ* 1970;**78**:311–29.
15. Darby MR, Karni E. Free competition and the optimal amount of fraud. *J Law Econ* 1973;**16**(1):67–88. https://doi.org/10.1086/466756.
16. World Commission on Environment and Development. Our common future: report of the world Commission on environment and development. *United Nations Comm* 1987;**4**:300. https://doi.org/10.1080/07488008808408783.
17. UNEP. *ABC of SCP. Clarifying Concepts on Sustainable Consumption and Production.* United Nations Environment Program; 2010. p. 1–59.
18. UNEP. *Paving the way for sustainable consumption and production: The Marrakech Process progress report.* UNEP; 2011. p. 104.
19. Eurostat. *Sustainable development in the European Union.* 2015. https://doi.org/10.2785/999711.
20. ESCAP. *Integrating the three dimensions of sustainable development: a framework and tools.* 2015. https://doi.org/10.1017/CBO9781107415324.004.
21. Morelli J. Environmental sustainability: a definition for environmental professionals. *J Environ Sustain* 2011;**1**:1–10. https://doi.org/10.14448/jes.01.0002.
22. Sogesid. *Sustainable Development.* WWW Document, http://www.sogesid.it/english_site/Sustainable_Development.html; 2017.
23. Lozano R. Envisioning sustainability three-dimensionally. *J Clean Prod* 2008;**16**:1838–46. https://doi.org/10.1016/j.jclepro.2008.02.008.

24. European Parliament, 2012. REGOLAMENTO (UE) N 1151/2012.
25. Grazia C, Hammoudi A, Malorgio G. Qualità e Sostenibilità delle produzioni Agroalimentari: Un modello Interpretativo delle Politiche Pubbliche e Strategie Private. In: Andreopoulou Z, Cesaretti GP, Misso R, editors. *Sostenibilità Dello Sviluppo E Dimensione Territoriale. Il Ruolo Dei Sistemi Regionali a Vocazione Rurale.* Milano, Italy: FrancoAngeli S.r.l; 2012. p. 87–133.
26. Bureau JC, Valceschini E. European food-labeling policy: successes and limitations. *J Food Distrib Res* 2003;**34**(3):70–6.
27. European Parliament, 2005. Regolamento (Ce) N. 396/2005 Del Parlamento Europeo E Del Consiglio. Gazz. Uff. dell ' Unione Eur. 16.
28. Council of the European Communities. REGULATION (EEC) No 315/93 laying down community procedures for contaminants in food. *Off J Eur Communities* 1993;**1993**:1–3.
29. European Commission. *REGOLAMENTO (CE) N. 1881/2006 DELLA COMMISSIONE del 19 dicembre 2006 che definisce i tenori massimi di alcuni contaminanti nei prodotti alimentari 2006;* 2006. p. 5–24.
30. The European Parliment and the Council of the European Union. *REGULATION (EU) No 1169/2011 OF THE EUROPEAN PARLIAMENT AND OF THE COUNCIL of 25 October 2011 on the provision of food information to consumers, amending Regulations (EC) No 1924/2006 and (EC) No 1925/2006 of the European Parliament and of the Council, an. Off J Eur Union* 2011;**7**:18–63. https://doi.org/2004R0726 - v. of 05.06.2013.
31. Fondazione Qualivita. *Made in Italy, Grana Padano DOP: Similari, situazione oltre il limite.* WWW Document, http://www.qualivita.it/news/made-in-italy-grana-padano-dop-similari-situazione-oltre-il-limite/; 2017.
32. Reganold JP, Wachter JM. Organic agriculture in the twenty-first century. *Nat Plants* 2016;**2**:15221. https://doi.org/10.1038/nplants.2015.221.
33. European Commission. *What is organic farming?* WWW Document, https://ec.europa.eu/agriculture/organic/organic-farming/what-is-organic-farming_en; 2018. (accessed 1.23.18).
34. INAO. *Label Rouge.* WWW Document, https://www.inao.gouv.fr/eng/Official-signs-identifying-quality-and-origin/Label-Rouge-Red-Label; 2018. (accessed 1.23.18).
35. Lamastra L, Suciu NA, Novelli E, Trevisan M. A new approach to assessing the water footprint of wine: an Italian case study. *Sci Total Environ* 2014;**490**:748–56. https://doi.org/10.1016/j.scitotenv.2014.05.063.
36. MATTM, 2016. VIVA "La Sostenibilità nella Vitivinicoltura in Italia" DISCIPLINARE ARIA PRODOTTO.
37. Viticolturasostenibile.org. *La gestione unica della sostenibilità.* WWW Document, http://www.viticolturasostenibile.org/News.aspx?news=352; 2017. (accessed 12.14.17).
38. GlobalGap. *A modular approach to integrated farm assurance.* WWW Document, https://www.globalgap.org/uk_en/for-producers/globalg.a.p./integrated-farm-assurance-ifa/; 2018. (accessed 1.25.18).
39. IFS. *IFS Food 6.1.* WWW Document, https://www.ifs-certification.com/index.php/en/standards/251-ifs-food-en; 2018.
40. 22000-tool. *What is ISO 22000?* WWW Document, http://www.22000-tools.com/what-is-iso-22000.html; 2018. (accessed 1.26.18).
41. SQF/Food Marketing Institute (Hrsg.), 2014. SQF code edition 7.2, A HACCP-based supplier assurance code for the food industry 66.
42. International EPD System. *International EPD system.* WWW Document, http://www.environdec.com/it/PCR/; 2017. (accessed 12.12.17).
43. International EPD System. *EPD database.* WWW Document, http://www.environdec.com/en/EPD-Search/?search_type=advanced&query=&country=Italy&category=6192&certEpd=false&deregEpd=false&sectorEPD=false&ecoPlatformEPD=false&en15804EPD=false; 2017. (accessed 12.14.17).

44. European Commission. *DOOR database*. WWW Document, http://ec.europa.eu/agriculture/quality/door/list.html;jsessionid=pL0hLqqLXhNmFQyFl1b24mY3t9dJQP flg3xbL2YphGT4k6zdWn34%21-370879141; 2018. (accessed 2.15.18).
45. European Commission. *E-Bacchus*. WWW Document, http://ec.europa.eu/agriculture/markets/wine/e-bacchus/index.cfm?event=resultsPEccgis&language=EN; 2018. (accessed 2.15.18).
46. Ceci-Renaud N, Clément M, Tarayoun T. Service de l'économie, de l'évaluation et de l'intégration du développement durable, In: Demeulenaere L, editor. Sous-direction de l'économie des ressources naturelles et des risques, Commissariat Général au Développement Durable, Tour Séquoia, La Défense cedex, ernr.seei. cgdd@developpement-durable.gouv.fr, www.developpement-durable.gouv.fr
47. Gruère G. A Characterisation of environmental labelling and information schemes, OECD Environment, Working Papers, No. 62. Paris: OECD Publishing; 2013. https://doi.org/10.1787/5k3z11hpdgq2-en.
48. Fischer C, Lyon TP. Competing environmental labels. *J Econ Manag Strateg* 2014;**23**:692–716. https://doi.org/10.1111/jems.12061.
49. Harbaugh R, Maxwell JW, Roussillon B. Label confusion: the Groucho effect of uncertain standards. *Manag Sci* 2011;**57**:1512–27. https://doi.org/10.1287/mnsc.1110.1412.
50. Vermeir I, Verbeke W. Sustainable food consumption: exploring the consumer "attitude—behavioral intention" gap. *J Agric Environ Ethics* 2006;**19**:169–94. https://doi.org/10.1007/s10806-005-5485-3.
51. Albright MB. *Organic foods are tastier and healthier, study finds*. WWW Document, http://theplate.nationalgeographic.com/2014/07/14/organic-foods-are-tastier-and-healthier-study-finds/; 2014.
52. OECD. *Greening household behaviour: the role of public policy*. Paris: OECD Publishing; 2011. https://doi.org/10.1787/9789264096875-en.
53. European Commission. Europeans' attitudes towards climate change. In: *Flash Eurobarometer 322*; 2009. 322, 134.
54. Centre de Recherche Pour l'étude et l'observation des Conditions de vie, Rapport No. 279, Département Conditions de vie et Aspirations. Etude Réalisée à la Demande de l'ADEME (Agence de l'Environnement et de la Maîtrise de l'Energie) Service Economie Prospective (SEP).
55. Council for Agriculture Science & technology. *Animal agriculture and global food supply 1999*; 1999. Task Force Report.
56. Nielsen. *The sustainability imperative: new insights on consumer expectations, Company Nielsen-Global Sustainability Report*; 2015.
57. Verbraucherzentrale E.V. Bundesverband, 2016. NACHHALTIGER LEBENSMITTELKONSUM—VON DER NISCHE IN DIE BREITE Positionspapier zum nachhaltigen Konsum im Lebensmittelbereich.
58. ISTAT, 2017. L'andamento dell'economia agricola—Anno 2016.

# CHAPTER FOUR

# The Role of Research, Communication, and Education for a Sustainable Use of Pesticides

## Maura Calliera*,1, Alba L'Astorina†

*Department for Sustainable Food Process, Dipartimento di Scienze e tecnologie alimentari per una filiera agro-alimentare sostenibile, Università Cattolica del Sacro Cuore, Piacenza, Italy
†Institute for Remote Sensing of Environment, National Research Council, Milano, Italy
1Corresponding author: e-mail address: maura.calliera@unicatt.it

## Contents

1. Introduction 110
2. Sustainable Agriculture and the Role of Sustainable Development European Policies on Pesticide Use Education 112
3. State of the Art: Lessons Learnt From Research Projects Aimed to Foster Education in Sustainable Use of Pesticide 114
4. Education and Communication for a More Integrated System of Knowledge 119
   4.1 Communication in Agriculture: New Challenges 121
   4.2 Effectiveness Teaching in Agriculture 122
   4.3 E-Learning Technologies: Strengths and Critical Points 123
   4.4 The Role of Higher Education in the Training of Future Farmers 124
5. The RRI Approach for a Sustainable Use of Pesticides in Agriculture 125
   5.1 Criteria 127
   5.2 Actions 127
6. Conclusions 129
References 130

## Abstract

Sustainable agriculture is not the result of a simple linear, one-way process that goes from the scientific production of knowledge to its production application, but the result of complex "systemic interactions" between different subjects and institutions involved in various ways in the production and dissemination of knowledge and its incorporation into innovative solutions applicable. This is way in the last few years agricultural sustainability and sustainable use of pesticide have been subject of important regulatory interventions and research projects that stimulated a search for new approaches and design systems for knowledge transfer. Innovation and networks have also gained increasing attention for their role in enhancing knowledge and interaction between stakeholders.

*Advances in Chemical Pollution, Environmental Management and Protection*, Volume 2
ISSN 2468-9289
http://dx.doi.org/10.1016/bs.apmp.2018.03.002
© 2018 Elsevier Inc.
All rights reserved.

In this chapter we will focus on:

- the contribute of policy in knowledge enhance and promotion of the adoption of innovation in agriculture,
- lessons learnt from innovative project that considers the diffusion of a culture of prevention and anticipation as the most appropriate tool to tackle the management of risks on health and the environment,
- the importance of communication and social interaction for sharing experiences and transmitting information, and
- the importance of a flexible education system and training tools in supporting a knowledge system perception-oriented, context-specific, more participative that involve different factors and different types of knowledge.

**Keywords:** Sustainable pesticide use education, Knowledge enhance, Responsible Research and Innovation, Training tools

## 1. INTRODUCTION

Many uncertainties question the way we produce, process, and consume food. We are challenged to reduce the climate footprint of our food systems and at the same time to consider the perspectives of different stakeholders involved in the sector.

Consumers and citizens are demanding more information on how food is produced, distributed, and consumed; European farming and agrofood systems in a globalized world look for competitiveness maintaining high-quality standards with low costs.

In this framework, the governance of agriculture must conciliate the right to a healthy, safe, secure, and sufficient food with the need to guarantee a sustainable model of production and consumption, where aspects such as ethics, culture, local knowledge, tradition, responsibility, innovation, and technology are all considered.

The impact of agriculture on many aspects of life, not only connected with food system, but also concerning the development of rural areas, bio-diversity, and ecosystem services, has stimulated a search for new approaches able to address complex issues that underlie this human activity. In particular adequate solutions are considered necessary to contrast negative impacts to human health and the environment connected with the use of chemicals.

Approaches are based on regulatory interventions on the use of pesticides in agriculture, which in the last few years have been object of consistent normative. One of the most important, the Directive EU 128/2009 on pesticide sustainable use,[1] requires Member States to develop specific measures to minimize environmental and also occupational exposure to pesticide and to promote alternative approaches or techniques, such as nonchemical alternatives.

Research and innovation is also given a crucial role in helping future agriculture security and competiveness not only based on technologies supporting an intensification of the agricultural production, but also increasing a knowledge system where the traditional "linear" model—from scientists to users—is gradually replaced by a more participative one which integrates knowledge production, adaptation, advice, and education involving all factors: from farmers, to researchers, policy makers, companies, and citizens.

In a scenario where pesticides in agriculture in following years and the search for alternative approaches or techniques, such as nonchemical ones, is still matter of debate, research projects address the importance of information, communication, and education of the farming community as key factors for the development of a sustainable management of agriculture. Lessons learnt from experiences described in these projects reveal that one of the biggest challenges is the cultural investment that takes into account both the global and local dimension and the specific socioeconomic context of farmers. While it is important to have professional figures able to manage the growing degree of complexity of a system characterized by an increasingly rapid production of knowledge, the removal of some bottlenecks, as the renewal and strengthening of technical skills of all factors involved and the delivery of practice-oriented research to end-users are also important.

Recent recommendations toward a Responsible Research and Innovation (RRI) a crosscutting theme of Horizon 2020[2] considering as central the necessity to anticipate and gain knowledge of possible consequences and building a collective capacity to respond to modern challenges[3] are part of this evolving scenario.

In the following paragraphs, we will focus on lessons learnt from innovative projects addressing the importance to involve farmers in the development of strategies toward a sustainable use of pesticides starting from their practices; on the role of both technoscientific research and policy to identify the drivers toward their adoption.

We will address issues such as the importance of communication, education, and training toward a responsible and sustainable management of chemicals in agriculture. We will then conclude with some recommendations for further research.

## 2. SUSTAINABLE AGRICULTURE AND THE ROLE OF SUSTAINABLE DEVELOPMENT EUROPEAN POLICIES ON PESTICIDE USE EDUCATION

Sustainable agriculture is a key objective of the European Union and a focus of its sustainable development policies.

An example of EU policy-oriented sustainability approach is the Sustainable Consumption and Production and Sustainable Industrial Policy Action Plan (SCP/SIP) presented on July 16, 2008,[4] an integrated set of actions and measures that represent the strategies of the European Commission for the improvement of the environmental performance of products and to increase the demand for more sustainable goods and production technologies.

In order to face with the public increasing concern for the potential adverse impact of pesticides on human health and the environment, the European Commission has introduced legislation aimed at reducing risks and impacts of pesticides in the use phase (Directive 2009/128/EC, hereinafter expressed as Directive).

In the Directive, Member States (MSs) are required to develop specific measures to minimize environmental and also occupational exposure to pesticide. The Directive aims to achieve a sustainable use of pesticides promoting the use of Integrated Pest Management (IPM) and of alternative approaches or techniques, such as nonchemical alternatives to pesticides.

MSs have drawn up National Action Plans to implement the range of actions set out in the Directive, ensuring that all the so-called professional users, distributors, and advisors—including operators, technicians, employed and self-employed in agriculture and other sectors—have access to "appropriate training" by bodies designated by the competent authorities,[5] highlighting the important role given to targeted training tools.

Another important legislative initiative is the Nitrate Directive or the most recently legislation dealing with pesticide production, licensing, use machine for pesticide application commonly called "Pesticide Package" which also include the Directive.

The effectiveness of these laws, however, can be reduced or slowed by several factors.

From the economic point of view, agriculture is comparable with other industrial activities and the implementation of sustainability legislation influences the production possibilities of the agricultural sector. The adoption of innovation at farm level is seen as a strategy or driver to reach the policy objectives. But different problems in addressing the challenge are still present. For example, the adoption of innovations, especially at farm level, often linked to the education levels and economics, is not a linear process.

Several barriers can limit the process, as for example the lack of resources, of technical expertise or of knowledge, cultural-related barriers as risk aversion, market characteristics considering many small players for which the transaction cost for innovation are rather high.[6] UC Davis Agricultural Sustainability Institute states that "making the transition to sustainable agriculture is a process. For farmers, the transition to sustainable agriculture normally requires a series of small, realistic steps. Family economics and personal goals influence how fast or how far participants can go in the transition."[7]

The above mentioned problems and challenges, if insufficiently addressed, make more difficult for public authorities of MSs to help stakeholders such as farmers to comply with rules set in the legislation.

In addition to the legislative aspect, it is also essential to analyze the role of policy in promoting the adoption of innovation in agriculture. Governments invest in research and development and is well known that research, innovation, and market are strictly correlated. New technologies, such as Spray Drift Reduction Technology (SDRT), or developed for agriculture/precision farming such as drones and apps coupled with geospatial information, ITC, etc., are some examples.

However, as reported in the Standing Committee on Agricultural Research (SCAR) reports, more research implies not necessarily adoption of innovation, especially in agriculture. Indeed, traditionally, for farmers such innovation activities are full of risks that have to be managed and this can hamper its adoption. Innovation is also linked to business and agricultural market provides too little R&D because agricultural producers *perceive the chance of success to be too low or the costs of innovations and experimentations too high*. In this framework additional policy activities are needed in order to push for a change or for the acquisition of an innovative product or services that have to be marketed.

Policy makers and a targeted policy innovation, context specific, could play a key role in stimulating the innovation process and reduce negative external effects linked to agricultural activities (as environmental pollution due to chemicals), focusing on the interaction between different stakeholders (researcher scientist, industry scientist, extension/advisors, farmers, etc.) and so promoting rural multifactor networks and with target fiscal measures, in addition of the environmental regulation (such as the directive).[8]

Success of policies will depend on one side to the promotion of programs and research aimed to the provision of novel, effective, and reliable approaches and tools to the farmers, on the other side on the capacity to involve and really connect farmers (who must manage them), to the world of research through adequate figures capable of translating research results.

## 3. STATE OF THE ART: LESSONS LEARNT FROM RESEARCH PROJECTS AIMED TO FOSTER EDUCATION IN SUSTAINABLE USE OF PESTICIDE

It is widely recognized that in all policy processes information and knowledge diffusion play a major role, but an important role is also played by the approach in which such processes are activated.

Information and training activities aiming at fostering a sustainable use of pesticides should concern all aspects related to the management of chemicals in the farms: risks associated with transport, storage, distribution, and disposal; control of the equipment for their distribution; strategies for the protection of water bodies and natural sites in the area; available technologies able to favor a sustainable use of pesticides in agriculture and to improve the environmental performance of farms.

The activities should not only inform about solutions that reduce the contamination of pesticides in agriculture but also promote the adoption of the prevention approach and among all factors. To this aim, the modality in which these activities are carried out is crucial, preferring customized meetings and participatory activities aimed at the active involvement of all the factors besides demonstration events in the field. Benefits of these approaches are to share quantitatively and qualitatively higher information, and to activate a process of empowerment of all the players in the supply chain.

Such methodological recommendations are based on a consolidated literature that considers the diffusion of a culture of prevention and anticipation as the most appropriate tool to tackle the management of risks on health and the environment in an ecosystemic and long-term perspective. This strategy presupposes an understanding of the context of the various factors operating in the supply chain, of the ways in which they behave in their daily practice, and of the many factors affecting their decision to assume or not assume a sustainable behavior.

Several research projects are moving toward this inclusive approach. Some innovative contributions, key lessons, and recommendations come from public national and international projects and from initiatives funded by private bodies in order to foster sustainable use of pesticides.

The EU FP7 BROWSE project (Bystanders, Residents, Operators and Workers Exposure models for plant protection products)[9] concluded at the end of 2015, had among the objectives that of contributing to the creation of materials for training and assessment of the risk perception resulting from exposure to plant protection products. BROWSE involved a large representation of stakeholders, such as farmers, consumers, industry, academy and research institutions, policy makers, NGOs, in a participatory consultation process in order to identify their opinions on: the topics considered as priorities for training, the factors that influence the exposure of pesticides, and the tools and formats deemed most appropriate for effective training and awareness. During the consultation, both the indications prescribed by the European Directive and the training material available or implemented in each country were evaluated. It emerged that much documentation has been produced in each Member State with a wide range of techniques described in several European projects, or within private initiatives, such as training activities promoted by European Crop Protection (ECPA).[10] However, despite the many resources available, these materials are often poorly understood by those concerned and rarely harmonized with each other.

The results of the survey also showed that there is a growing interest in the adoption of more "modern" communication approaches—experimental, demonstrative, and participatory—and of more appropriate techniques, with a clear preference for material in audio–video format. In order to effectively inform, they recommend then to use a combination of training techniques such as lessons and group discussions, followed by practical demonstrations, which allow "learning" through practice and promote the understanding of the issues addressed.[11]

Also focused on the methodology of information introducing Information and Communication Technologies (ICT) tools is the Erasmus + SUPTRAINING[12] project, aimed at building an e-learning platform aimed at providing the contents necessary to obtain the mandatory certification for the purchase and use of plant protection products. As one of the products of this project, the Guidelines for the sustainable use of plant protection products are included, in the context of the application of the Directive, which provides that "Member States shall take steps to ensure that the handling, storage and handling of packaging and inventories do not pose a danger to human health and the environment."

In this context a series of actions are foreseen, including the "development of safe procedures for the storage and handling of plant protection products, the preparation of the phytoiatric mixture, the washing of containers and machinery after treatment, the disposal of water waste and packaging."

The Life project EcoPest[13] was a pilot-scale demonstration project ended in 2012 aiming at the protection of a vulnerable inland water ecosystem from pollution caused by the excessive use of pesticides. Funded by the European Commission within the framework of the Life+ Environment Policy and Governance Program, EcoPest involved farmers, agronomists, agricultural research institutes, agricultural and crop protection industry/companies, academia, public bodies, and policy makers toward the definition of a complete scheme for prevention from excessive and improper pesticide inputs and protection of the environment and human health.

The EU FP7-KBBE PRO AKIS (Prospects for farmers' support: Advisory services in European AKIS)[14] project aimed to explore and systematically describe the various European Agricultural Knowledge and Innovation Systems (AKIS) schemes. Work began by creating an inventory of AKIS organizations, institutions, and their linkages in 27 MSs. Four case studies were conducted in seven MSs focusing on small-scale farmers' access to relevant and reliable knowledge, services bridging scientific research topics and farmers' demands, and appropriate support for diverse rural factors that form networks around innovations in agricultural and rural areas. The studies in this project were taken into account in the report of the Standing Committee on Agricultural Research (SCAR) on Agricultural Knowledge and Innovation Systems.

In the TRAINAGRO project,[15] just started and focused on some Lombard agricultural areas, a SWOT analysis is used in order to gather data and address a more efficient communication strategy involving all factors. The SWOT methodology allows analyzing the theme from four different points of view: the identified strengths help to understand what to focus on and how to take decisions; weaknesses helps to understand where to intervene and how, opportunities where to invest, and, if identified, how to transform threats (risks) into opportunities. An investigation activity, like that made in TRAINAGRO, in addition to producing useful data for the project, also allows to overcome the gap between the scientific world and agricultural operators, and enhances the sharing of knowledge, experiences, and possible solutions among the participants. In addition to SWOT, other participatory methods such as Open Space Technology, Metaplan, and Wordcafé will be used to ensure maximum interaction among partners as already shown in analogous projects.[16]

Further projects funded by private bodies have looked for new solutions to sustainable use of pesticides. Among them, the training activities developed by ECPA[10] about the pesticide storage aspects, the safe disposal of container, fake products, and illegal use and, with the focus on water protection, with guidelines produced in the TOPPS PROWADIS[17] project that started in 2012 with the aim to reduce pesticide entries into surface water from applications in the field. It builds on the TOPPS-Life project (focusing on point sources pollution) and has its focus on aspects related to runoff/erosion and spray drift.

Networks have also gained increasing attention for their role in enhancing knowledge and interaction between stakeholders. Generally they are context specific, take into account the diversity of knowledge, focus on concrete objectives of end-users, and practice application with the engagement of various factors as cocreators of solutions and for this are element for funding.[18] Below some interesting examples.

The overall objective of Eu FP7-KBBE SOLINSA[19] (Agricultural Knowledge Systems in Transition: towards a more effective and efficient support of Learning and Innovation Networks for Sustainable Agriculture) project is to identify effective and efficient approaches for the support of successful LINSA (Learning and Innovation Networks for Sustainable Agriculture) as drivers of transition toward Agricultural Innovation Systems for sustainable agriculture and rural development. The project identified factors that promoted or held back existing agricultural knowledge systems and innovation policies. Traditional understanding of knowledge transfer has to shift to a conception of knowledge exchange between equal partners. Agricultural knowledge systems have to widen their scope and open up for a creative exchange of innovators for sustainable agriculture and rural development from different fields. In this situation, transition partners emerge as new kind of factors, with particular roles and functions. These are various kinds of networkers, facilitators, participatory researchers, and boundary persons who are experts who engage with projects such as LINSA in joint learning and innovation for sustainability.

The Eu FP7-KBBE FarmPath[20] (Farming transitions: pathways towards regional sustainability of agriculture in Europe) proposes that increasing sustainability in agriculture is best addressed by enabling flexible combinations of farming models, which vary to reflect the specific opportunity sets embedded in regional culture, agricultural capability, diversification potential, ecology, and historic ownership and governance structures. Using literature reviews and EU statistical databases, researchers also found that the presence of young farmers is related to the size of the agricultural sector, as well as profitability. Researchers also found that agricultural innovation often comes from outside the agricultural industry, and suggest that cross-sectorial engagement would improve this situation. Twenty-one sustainable agriculture case studies have been completed in various EU Member States and a network called the National Stakeholder Partnership Groups was created to guide future work.

Interesting initiatives are also promoted by the Endure network that aims to brings together some of Europe's leading agricultural research, teaching and extension institutes with a special interest in IPM within the general context of environmentally friendly and sustainable agriculture.

Born as Network of Excellence funded by the European Commission in 2007 at the end of the funding period in 2010, partners committed to keeping ENDURE going as a self-funded European Research Group and "continue to operate at the heart of IPM research and extension, working alongside others in this dynamic and challenging field to identify those areas where further research is needed, possible funding sources and providing expert assistance, both nationally and at a European level, to ensure IPM continues to develop as a sustainable, cost-effective and achievable way of contributing to food security and a better environment" (from http://www.endure-network.eu/).

The Magic Nexus Project "Moving Towards Adaptive Governance in Complexity: Informing Nexus Security" (MAGIC)[21] aims at facing new challenges for the governance of sustainability, implementing policies involving the Nexus between water, food, energy, and land use. The goal of MAGIC is to transform Nexus from a shorthand to signify the complexity of the relationship between water, soils energy, and climate into a set of relationship over identified factors which can be systematically used to explore this complexity. This implies integrating into the analysis of social challenges and stakeholders perceptions related to the climate–water–food–energy Nexus. The project opens dialogue spaces, enacts dissemination strategies, and develops mixed qualitative–quantitative tools in the context of a community-building exercise transcending mechanistic scientist–policy maker separation but taking full advantage of the rich spectrum of factors and institutions active in the Nexus.

All these projects share a participative and inclusive approach considered as necessary in all phases of the relationship with stakeholders, starting from a deep understanding of the behavior of the various factors, to more interactive communication and demonstration strategies, up to training activities that overcome the traditional top-down approach "teacher–student" and consider local knowledge as an important key for the transition toward sustainability.

## 4. EDUCATION AND COMMUNICATION FOR A MORE INTEGRATED SYSTEM OF KNOWLEDGE

Today the line between education and training could be very thin. Commonly, education is linked to the acquisition of knowledge while training of skills. There are many definitions in the literature concerning training in agriculture and several approaches, methods, and techniques. Despite the great efforts posed by the Directive, *"training in agriculture cannot be considered only in terms of an educational process;* it must also be taken as part of a wider socioeconomic environment in which complex, worldwide factors interact with each other."[22]

In consolidated literature, produced as results of the project previously cited, training in agriculture should be a bottom-up rather than a top-down process (from science to implementation). Agriculture training should be implemented taking into account the various roles and responsibilities, and should ensure that users, distributors, and advisors acquire sufficient knowledge regarding the subjects listed in Annex 1 of the above cited Directive.

However, as also stressed in the Horizon 2020 Work Program 2016–2017, in 2010, 71% of European farm managers are operating on the basis of practical experience only. Education levels vary greatly depending on country, farm manager's age and gender, or farm structures, and this can hamper innovation[23] and confirms the connection between farmers socioeconomic status (income, education, age, gender, access to information, or availability of extension service) and the importance of considering all this in actions aimed at promoting farmers' sustainable behavior in the management and use of pesticides.

The educational approach should take into account that pesticides are used in very complex environments, which are heterogeneous in terms of conditions of use, social conditions, and cultural backgrounds. Data from the 2017 Report on young farmers of EU DG Agriculture and Rural Development confirm that there are roughly 3.2 million farmers older than 65 years in the EU and in 2013 the majority of farm managers (68.2%) learned their profession through practical experience only.[24] The Report highlights the importance to improve the educational status of farmers and to provide access to professional training in line with the Directive.

At present, the designed system of knowledge transfer does not reduce complexity, but is instead perceived as inevitable and complicate, programmed or designed to be an expensive control mechanism that transfers the responsibility instead to be designed to promote the acquisition of additional knowledge, improve understanding of complex and interrelated problems, induce to making conscious choices and tracking progress, and motivate factors toward a sustainable production.

In order to address all the knowledge required, pluralistic research is needed, considering the inter- and transdisciplinary and value-based nature of sustainability itself and the magnitude of the challenges. The role of social interaction has to be taken into account and evaluated, to achieve a successful transition vs sustainability approach that rely on activities that are essentially relational in nature and require the creation and maintenance of a connection between one or more factors.[25]

Commonly it is argued that engagement in unsafe pesticide use and disposal practices is the result of a lack of knowledge and misperceptions of the risks associated with pesticides among operators and workers. Several published studies show that in developing countries high illiteracy rates contribute to farmers' (and farmworkers') difficulties in understanding and following pesticide use instructions and safety advice.[26] But despite this evidence, a professional pesticide user is subject to biases and heuristics in thought processes, has a set of preferences, and is ultimately responsible for his/her actions. The operator decides the time of the treatment and the way to operate.

Several studies[27–29] confirm that operator behavior can influence environmental and human exposure at all stages of pesticide management, behavior is very much linked to the perceptions and personal attitude toward risks, and that education level is not necessarily related with a more correct risk perception of pesticides use.

Indeed, despite the level of education or care, and good training, in many situations observations show that awareness of the problem is not always supported by a fully rational behavior.

These results substantiate issues already highlighted in a previous study carried out in Chiapas in Southern Mexico among people involved in pesticide use in tomato and banana cultivation and confirm that the issue is common for both developing and developed countries, with a generally higher level of education.

The level of formal knowledge is not the only explaining factor for how much of the technical recommendations are followed by factors. Existing training concepts will fail as they are based on the classical expert–trainee division if the model foresees that pesticide users are *just recipients of formal knowledge and have no knowledge, skills, or experiences to tap into*.[30] The above mentioned reflection involves examining the knowledge base and theories in use that inform the practitioners to which such knowledge and theories are disseminated. It is necessary to identify the so-called training need that is "the gap between what is and what should be in terms of incumbents' knowledge, skills, attitudes, and behaviour for a particular situation at one point in time."[20]

There is the need to push for an educative approach linked to the ability to generate new knowledge overcoming traditional science teaching made of concepts, often not related with the real world[31] able to develop new form or mode of relationships.

## 4.1 Communication in Agriculture: New Challenges

Today a relevant problem affects the relationship between the ever-increasing diffusion of communication channels and the possibility of obtaining really reliable, verified, correct, and functional information for daily practice on the issues that daily closely affect the agriculture sector.

It has been argued that the increase in tools and means of communication, from Internet to the world of social networks, does not always favor the production of good communication.

In the traditional communication processes for rural development, the receptor was only a passive subject that allowed to reach or satisfy numerical objectives.[32,33]

Many development projects supported by the United Nations, FAO, and World Bank are a clear example of this process. The finding that most of the supposed beneficiaries of these projects did not receive a real benefit, caused alarm and concern, prompting these institutions to seek alternative solutions based on some trends that are now widely shared and which concern:

**(a)** User orientation: The old practice of using the same technical messages for all farmers, using the same methodology of disclosure, is gradually replaced by client-oriented approaches. In fact, there are many needs of elderly farmers, farmers interested in the market, young people, and rural women.

**(b)** Participated agricultural dissemination—The tendency to involve farmers in making decisions that affect them has led to the dissemination of new methods and to the development of participatory tools.

From the literature on the subject, it is clear that it is essential to adapt to the ways in which the recipient of the message processes the information and knows its level of knowledge, so as to start from it in the process to develop it. The receptor, previously passive, is activated starting to collaborate in the numerous options that opens the communication process. According to Manuel Calvelo Rios (awarded by the FAO with the Sen award as best expert for his contribution to the Communication for rural development) there is real communication if and only if the messages that are interchanged are the product of a joint work, as the etymology of the term indicates. Indeed communicating means "doing together."

In the framework of pesticide use, it has been widely recognized that their sustainable use can only be successful with involvement of farmers and other stakeholders.[26] Nevertheless, communication and interactions with farmers are complex and defining the ideal form for this engagement is not a simple task. The educational landscape has to change and must become more responsive to allow for integrated training program solutions that take present-day and future challenges into account to enable people to be less individualist, better informed, and more autonomous in their choices and that better link knowledge and action to the general philosophy of sustainability.

Social interaction, as suggested by the Standing Committee on Agricultural Research (SCAR),[34] could be the engine for learning and innovation, toward a change of innovation models: "from top-down models to network models, from generalist extension structures (for example, 'crop management' departments) to problem-specific structures (for example, 'participatory plant breeding' networks), and will allow direct interaction between farmers, researchers, extension workers, consumers and civil society organisations. Sustainable development and Innovation policies can support the restructuring process by raising the level of digital literacy, encouraging collective development projects with explicit Internet-based experiments, and introducing methods of monitoring and evaluation of learning processes."

## 4.2 Effectiveness Teaching in Agriculture

Merging the outcomes of target European and national research projects, it is possible to conclude that the delivery of operator training for a sustainable use of PPPs needs to be done using a mix of appropriate techniques including the use of modern communication technologies and using audio–video material as preferable format. Furthermore, as explained before, it seems evident that a shift from a formal training on pesticide use that supposes lack of knowledge by farmers to a training system more perception-oriented and context-specific is needed to achieve the pesticide Sustainable Use Directive goals.

Indeed, priorities for training differ according to target groups and the means of delivering the training need to be sufficiently diverse using flexible tools to be able to address all differences in knowledge and resources.

Very often different partners (extensors, advisors, experts, distributors, etc.) could have "a role" to play in the knowledge generation process, as each group brings an important perspective. It is important therefore to leverage both on training and innovative teaching methods but also on all aspects that connect science and society and their respective expectations about adopting the results of a "responsible research." The extent of scientific advances, ITC, and communications allows shorter assimilation periods. The increasing concern with environmental protection and the preservation of natural resources makes teaching on subjects like pesticide use more pressing. Thus, approaches, methods, and techniques in training need to be constantly updated, taking into account current trends and influences if they are to meet needs and realities of societies to achieve a sustainable development. Also teaching effectiveness has been described in several ways among educational researchers but good training is not only dependent on teaching strategies but also, especially in the case of farmers, on individual needs and adequacy of the content.

We remark that "the ability to communicate effectively with farmers in a way and using methods more reflective of the real-world practice is one of the main responsibilities of agricultural education trainer."[35] Rather than purely academic, education on sustainable use of pesticide must lead to the achievement of a "vision" capable of supplying the necessary transformational processes and should be supported by "*soft skills*" like learning to learn, think critically, to work together, research and apply new knowledge, make socially negotiated and shared decisions.

## 4.3 E-Learning Technologies: Strengths and Critical Points

The already mentioned SUPTRAINING project highlighted benefits from using e-learning technologies to train operators for a sustainable use of PPPs.

As reported in the FAO report on e-learning e-learning solutions offer effective instructional methods, such as practicing with associated feedback, combining collaboration activities with self-paced study, and personalizing learning paths based on learners' needs and using simulation and games. Thus, the use of e-learning solutions could guarantee greater flexibility, increased levels of on-the-job training, better quality in the training especially across multiple locations, more customization of the training to suit specific needs, and cost savings. The cost savings include reduced travel time or time off to complete the off-the-job components of the training.[36] However, it is important to consider that e-learning solutions are not ideal for all purposes and it is unlikely that these will completely replace classroom training in an organization. Indeed, a major barrier to e-learning is the challenge of changing mind-sets that are still locked into the traditional models of training delivery.

On the one hand, the majority of teaching/training staff are still adjusting to the reality that organizations want more flexible and engaging learning opportunities, packaged to suit their needs. On the other hand, students expect to use the technologies they are accustomed to and comfortable with. Thus, the most cost-effective application of e-learning may be to complement conventional training in order to reach as many learners as possible[37] Furthermore, it is clear that e-learning technologies are not useful tout court but represent an approach that will need to be developed, taking into account the specific context.

## 4.4 The Role of Higher Education in the Training of Future Farmers

With a view to prevention and anticipation another aspect that should be promoted is the involvement in policies promoting the sustainable use of pesticides in the agricultural sector of the school world.

Throughout the supply chain that goes from training to production in the agricultural field, students are the subjects most interested in understanding in advance the problems that will be faced in taking on the various professions in a production sector increasingly oriented toward sustainable development.

Many agricultural schools host within their boundaries agricultural farms, sometimes even of a certain size, which play a fundamental role for the practical application of the theoretical topics treated in the classroom and for the acquisition of specific skills in the agricultural sector. Involving students and teachers of agricultural schools in the policies of sustainable use of pesticides allows these factors to deal with the past (lesson learnt), the present (innovative tools at the moment available on the market), and to project themselves toward the future, when new operators will play the various roles in the agricultural field.

The involvement of schools for this purpose should be very precocious, beginning during the information and demonstration activities in the agricultural field, where students and teachers become a sort of "privileged witness" of the way in which the issue of sustainable use of pesticides is experimented and faced on the field by farmers, of the range of scientific and technological solutions offered today by the research world, with the possibility of combining the needs of the ones and the solutions offered by others.

On the other hand the schools, projecting toward the professional future, could be involved in supporting new communication and relationship ideas with the agricultural operators, and become promoters of new innovative solutions that also go toward the use of pesticides as already explored in other public initiatives involving students in the research and innovation process.[38]

## 5. THE RRI APPROACH FOR A SUSTAINABLE USE OF PESTICIDES IN AGRICULTURE

The fragmentation and the size of farms in Europe, that are often small or medium-small and with sometimes low incomes, are among the main weaknesses to tackle complex problems that often require advanced and shared solutions.

Reports done in order to evaluate the level of implementation of development policy (including the recent DG SANCO overview report on sustainable pesticide use)[39] highlight incorrect use of fertilizers and/or pesticides potentially dangerous to the environment (surface and ground water) and to the health of operators. They also highlight that farmers almost never use models based on agrometeorological and territorial conditions; the lack of general decision support systems; and the lack of environmental shared standard of quality on the territory.

To meet the challenges mentioned above, site-specific measurement tools for environmental impact generated by the type of agrochemical product used are needed, by the spraying systems used, by the quantity and frequency of treatment.

The sharing and transfer of information between companies producing or distributing pesticides and other subjects as researchers of scientific world, consultants, etc., is an opportunity to identify together and undertake synergistic actions aimed at environmental, social, and economic sustainability with the dual effect of protecting the territory and strengthen the image of the entire sector also according to the expectations of civil society.

In the last years, the European Commission (EC) has promoted an approach that seeks to anticipate and assess potential implications and societal expectations with regard to research and innovation, with the aim to foster the design of inclusive and sustainable research and innovation.[40] The approach, called Responsible Research and Innovation (with its acronym RRI), has become a crosscutting theme of Horizon 2020 the biggest European research and innovation program that funds "sustainable solutions to the challenges of the 21st century, such as global warming, energy, water and food, ageing societies, public health, pandemics and security."[41]

The approach of RRI considers as central the necessity to anticipate and gain knowledge of possible consequences and building a collective capacity to respond to them.[3]

In order to reach this goal, theory and practices of RRI recommend all societal factors—researchers, citizens, policy makers, business, third sector organizations, etc.—to work together during the whole research and innovation process in order to "better align both the process and its outcomes with the values, needs and expectations of society," in the prospective of a reciprocal responsibility.

An "ideal" RRI process should help scientists to identify four dimensions in their activities: anticipation, reflexivity, inclusion, and responsiveness.[3] This means that they should be able to understand (anticipate) how the current processes will effect and define future needs; examine and reflect on actions and consequent effects concerning all aspects of research and innovation: from daily routines, planning assumptions, and personal interactions, all the way up to institutional values and strategies.

A wide range of stakeholders should be involved in an inclusive way throughout the whole research process, in order to generate diverse perspectives and expertise. Finally, activities should be flexible and open to adapt existing organizational structures in response to evolving environments, values, and insights. In this formulation, RRI would go beyond the traditional risk assessment, concentrated to balance risks and benefits connected to the introduction of a technological innovation in the market.

This approach helps to manage risk perception of different public in function of the results of risk assessment and regulation and to manage the societal conflicts in case of uncertainty prevalence or high stakes.

Integration of the RRI dimensions at various stages in the research concerning the use of pesticides could help to overcome some barriers but should take into account several criteria and actions as illustrated in the paragraph below.

## 5.1 Criteria

- Incorporate an adequate analysis of the background (societal role, diverging problem definitions, etc.), addressing ethical, legal, social, and/or environmental aspects, envisioning plausible futures, and facilitating deliberation on values, perceptions, needs, and interests, now and in the future.
- Provide and collect data for science considering real world and context-specific research questions.
- Includes knowledge about best practices, heuristic, and terminology, but also identifying knowledge gaps.
- Making knowledge available for scientific purposes requires practitioners to share their knowledge, and be available for collaboration with researchers.
- Be Open and Transparent. Indeed the separation of the science-driven research activities and the management of the results provided by scientist could be responsible for a society science-hostile trend "as the role of science in issues of interest to all or many sections of civil society grows, so does the impact of any misunderstanding or values-based conflict relating to that science." Practitioners and researchers often come from very different backgrounds, use different terminology, and have different interests and priorities. The research should communicate appropriate, honest, and clear information about processes, roles, content, and results and give transparent public information in order to facilitate fast innovation, constructive collaboration among peers, and productive dialogue with civil society. Communication strategy can leverage on benefit, taking care to avoid the ideological component but, in opposite, enhancing the positive effect of behaviors and practice that generates strong motivation and high levels of involvement.

## 5.2 Actions

- Collect. There is a clear need to find a way to collect and integrate the available material as these are often unknown to the interested parties and not well harmonized among MSs in order to ensure farmers that all available knowledge is accessible.
- Looking "at the field" and experiential approach. Although pesticide Sustainable Use Directive provisions could be implemented, some gaps need to be filled in and the behavior of farmers needs to be nudged. Classical operator training as designed now appears critical to achieve a sustainable use of pesticide, understanding good practices, and improving behavior

gaps (i.e., following label instructions and use of Pest Protection Equipment (PPE), risk management) (from EcoPest, 2012).

In the pesticide risk evaluation process, mitigation measures are considered to reduce risk and the authorization is bound to the so-called good agricultural practices. However, it is recognized that risks have social and psychological dimensions[42] and are shaped by values, beliefs, attitudes, political systems, and cultural factors. It was noticed that due to the high geographical variability, farm structure, and cropping pattern, farmers need different sets of tools depending on their attitude toward risk. In this sense, a farmer's behavior could compromise the effectiveness of risk assessment. For this reason, a new sort of holistic approach, which takes also into account these variabilities, would help operators to understand the importance to well manage pesticide and to make the most appropriate choices leading to minor environmental and human risk without compromising the profitability of agricultural production and food standards.

In this framework with experiential approach the trainee becomes active and influences the training process. Potentially the experimental approach has a flexible structure that can be better adapted to different kinds of situations. Indeed, "in the experimental approaches methods, technics and other elements are jointly determined by trainers and trainee." Thus, trainers are not simply transmitting knowledge but primarily serve or operate as moderators, facilitators, catalysts, or resources persons.

Unlike the traditional approach, experiential training emphasizes real or simulated situations in which the trainees will eventually operate. Farmers are generally very busy and to initiate participatory training it is therefore very important to highlight the benefits this kind of training provides. If the participants of a participatory training session can see the added value it offers, it will keep them motivated and interested.[43]

- Link environment and farmers and Demo farming participatory events. It is also important to improve the "farmer image and proudness" and restore public confidence in farmers' activities, the sensibility of farmers to social pressure, and their investments and commitments to pro-environmental actions.[44] Demo farming will share and coordinate experiences, tools, and best practices. The use of sharing tools and the promotion of concrete initiatives could facilitate the spread of sustainability practices and, at the same time, to promote a coherent and coordinate approach that can facilitate the birth of useful collaborations and tangible synergies. Sharing existing knowledge and disseminating new technical solutions, indeed, are necessary to ensure that all the players can have the tools to improve the environmental, economic, and social sustainability. *Farmers and researchers can make more use of demonstration fields and field days to create an understanding for food production. The real risk* vs. *the perceived risks of consumers should be explained by science in "easy-to-understand" messages. All possible communication channels and media should be engaged in producing clear and simple messages for the general public* (Strategic research agenda for IPM in Europe-ERA Net C-IPM November 2016).

## 6. CONCLUSIONS

Many steps and investments have already been made both by the farmers and by researchers toward the diffusion of a culture of sustainable use of chemicals in agriculture and much can be still done starting from lessons learnt. An important role is played, however, by an appropriate policy of innovation that properly takes into account the barriers (cultural, socioeconomic, pressures of the market, context specific, etc.) and supports adequate educational system that takes the different level of knowledge of farmers in different contexts and the importance of networks in sharing experiences and transmitting information.

As a conclusion, in order to enhance the implementation and development of sustainable use of chemicals, following needs should be taken into account:

- Need to collect. A significant amount of training material is already available in a wide combination of technics (e.g., paper based materials, web software, videos).
- Need to complete. Still not every MS has applied the Directive requirements in terms of training and awareness raising and thus there is a different availability of training and awareness raising material in each EU MSs.
- Need to update. There is an increasing demand of a modern approach (especially referring to the so-called experimental or participative approach). Some successful examples have been identified.

In addition, according to recent studies, innovative performances, and therefore productivity in agriculture, are not the result of a simple linear, one-way process that goes from the scientific production of knowledge (research) to its production application, but the result of complex "systemic interactions" between different subjects and institutions involved in various ways in the production and dissemination of knowledge and its incorporation into innovative solutions applicable.

From the research, education, and extension triangle, we have moved on to the network of interacting heterogeneous and dynamic subjects that go beyond the traditional boundaries of the system. The innovation system provides, alongside the technological dimension, the social and environmental one, for which other factors such as consumers, citizens, opinion movements, the agricultural-rural community, the sectors of transformation and marketing of products from agriculture and, obviously, the whole system of transmission of knowledge guaranteed by institutions of higher education (secondary or university students).

A fundamental challenge is also to eliminate the obstacles related to the transmission of results of research oriented toward the final users (European Commission 2016).

The promotion of solutions with a real impact therefore requires an interdisciplinary approach and the integration of different types of knowledge. In fact, the agricultural operator is only one of the factors in the whole process, even if it is clear that he plays a fundamental role in determining the level of contamination; in fact, he decides, for example, the type and time of treatment and the methods of application, thus going to affect the final result.

Education and training aimed to improved skill, continuous updating, and public participation in the planning process are considered essential to achieve the objective of a sustainable use of chemicals in agriculture.

## REFERENCES

1. EU 128/EC, 2009. Directive of the European Parliament and of the council establishing a framework for community action to achieve the sustainable use of pesticides. OJ L 309/71 24.11.2009.
2. SiS.net. *RRI opportunities in Horizon 2020. Science with and for Society relevant topics in the Horizon 2020 work programme 2016–17*; 2016. 226 pp.
3. Stilgoe J, Owen R, Macnaghten P. Developing a framework for responsible innovation. *Res Policy* 2013;**42**(9):1568–80.
4. http://eur-lex.europa.eu/legal-content/EN/TXT/?uri=CELEX:52008DC0397.
5. EC 128/2009 Art. 5.
6. Hadjimanolis A. Barriers to innovation for SMEs in a small, less-developed country (Cyprus). *Technovation* 1999;**19**:561–70. from EU SCAR 2013.
7. Gail Feenstra, http://asi.ucdavis.edu/programs/sarep/about/what-is-sustainable-agriculture.
8. Agricultural knowledge and innovation systems (AKIS) towards the future—a foresight paper, Brussels. ISBN 978-92-79-54548-1. https://doi.org/10.2777/324117. ISSN 1831-9424 KI-NA-27-692-EN-N.
9. BROWSE project, FP7 Theme: environment (including climate change), bystanders, residents, operators and workers exposure models for plant protection products, grant agreement no: 265307, Project start date: 1st January 2011, Project end date: 31st December 2013, www.browseproject.eu.
10. http://www.ecpa.eu.

11. Sacchettini G, Calliera M, Marchis A, Lamastra L, Capri E. The stakeholder-consultation process in developing training and awareness-raising material within the framework of the EU directive on sustainable use of pesticides: the case of the EU-project browse. *Sci Total Environ* 2012;**438**:278–85.
12. SUPTRAINING project, Programme Erasmus + key action 2 "Strategic Partnership", development of an e-learning platform for the sustainable use of pesticides, Project reference: 2014-1-FR01-KA202-008754, www.sup-training.eu.
13. LIFE 07 ENV/GR/00266 EcoPest (www.ecopest.gr).
14. PRO AKIS (Prospects for farmers' support: advisory services in European AKIS) KBBE.2012.1.4-07—agricultural knowledge and innovation systems for an inclusive Europe Project ID: 311994.
15. www.trainagro.iambientale.it.
16. L'Astorina A, Tomasoni I, Basonie A, Carrara P. Beyond the dissemination of Earth Observation research: stakeholders' and users' involvement in project co-design. *J Sci Commun* 2015;**14**(03):C03. Special Issue, http://jcom.sissa.it/archive/14/03/JCOM_1403_2015_C01/JCOM_1403_2015_C03.
17. www.topps-life.org.
18. EU SCAR (2015), Agricultural knowledge and innovation systems (AKIS) towards the future—a foresight paper, BRUSSELS. ISBN 978-92-79-54548-1. https://doi.org/10.2777/324117. ISSN 1831-9424 KI-NA-27-692-EN-N.
19. SOLINSA (Agricultural knowledge systems in transition: towards a more effective and efficient support of learning and innovation networks for sustainable agriculture). KBBE.2010.1.4-04—knowledge systems for farming in the context of sustainable rural development—project ID: 266306.
20. FarmPath, (Farming transitions: pathways towards regional sustainability of agriculture in Europe) KBBE.2010.1.4-03—assessment of transition pathways to sustainable agriculture and social and technological innovation needs—call: FP7-KBBE-2010-4 ID 265394.
21. http://magic-nexus.eu/nexus-times.
22. Halim A, Mozhar A. Training and professional development. In: Swanson BE, Bentz RP, Sofranko AJ, editors. *Improving agricultural extension. A reference manual.* Rome: FAO; 1997.
23. HORIZON 2020. Work Programme 2016–2017 Section 9—food security, sustainable agriculture and forestry, marine and maritime and inland water research and the bioeconomy.
24. EU Agricultural and Farm Economics Briefs n15/Oct 2017. Young farmers in the EU—structural and economic characteristics. DG Agriculture and Rural Development, Unit Farm Economics. https://ec.europa.eu/agriculture/sites/agriculture/files/rural-area-economics/briefs/pdf/015_en.pdf.
25. Frans Hermans, The potential of exponential random graph models and stochastic actors oriented models for transition studies, The 2015 annual conference of the EU-SPRI Forum Innovation policies for economic and social transitions: developing strategies for knowledge, practices and organizations.
26. Damalas C, Eleftherohorinos I. Pesticide exposure, safety issues, and risk assessment indicators. *Int J Environ Res Public Health* 2011;**8**(5):1402–19. Remoundou K, Brennan M, Hart A, Frewer L. Pesticide risk perceptions, knowledge, and attitudes of operators, workers, and residents: a review of the literature. *Hum Ecol Risk Assess* 2014;**20**(4):1113–38.
27. Damalas CA, Abdollahzadeh G. Farmers' use of personal protective equipment during handling of plant protection products: determinants of implementation. *Sci Total Environ* 2016;**571**:730–6. https://doi.org/10.1016/j.scitotenv.2016.07.042.
28. Sacchettini G, Calliera M. Link practical-oriented research and education: new training tools for a sustainable use of plant protection products. *Sci Total Environ* 2016;**579**:972–7. https://doi.org/10.1016/j.scitotenv.2016.10.166.

29. Calliera M, Berta F, Galassi T, Mazzini F, Rossi R, Bassi R, et al. Enhance knowledge on sustainable use of plant protection products within the framework of the sustainable use directive. *Pest Manag Sci* 2013;**69**(8):883–8.

30. Ríos-González A, Jansen K, Javier Sánchez-Pérez H. Pesticide risk perceptions and the differences between farmers and extensionists: towards a knowledge-in-context model. *Environ Res* 2013;**124**:43–53.

31. Hayden K, Ouyang Y, Scinski L, Olszewski B, Bielefeldt T. Increasing student interest and attitudes in stem: professional development and activities to engage and inspire learners. *Contem Issues Tech Teacher Edu* 2011;**11**(1):47–69.

32. Toschi L, et al. In: Maggioli, editor. *La comunicazione sostenibile per lo sviluppo rurale*; 2016.

33. Liano Angeli La comunicazione per lo sviluppo rurale nei progetti Fao-Italia Agriregionieuropa anno 6 n°22, Set 2010.

34. EU SCAR (2013), Agricultural knowledge and innovation systems towards 2020—an orientation paper on linking innovation and research, Brussels. ISBN 978-92-79-32766-7. https://doi.org/10.2777/3418.

35. Shinn YH. *Teaching strategies, their use and effectiveness as perceived by teachers of agriculture: a national study*; 1997. p. 12244. Retrospective Theses and Dissertations, http://lib.dr.iastate.edu/rtd/12244.

36. Callan, V., 2009. How organisations are using e-learning to support national training initiatives, Australian flexible learning network. http://flexiblelearning.net.au/wpcontent/uploads/E-learning_and_National_Initiatives_Report.pdf [accessed 20 October2016].

37. FAO, 2011. E-learning methodologies—a guide for designing and developing e-learning courses. (ISBN 978-92-5-107097-0).

38. La Ricerca va a Scuola (http://space4agri.irea.cnr.it/it/scuola) L'Astorina, I. Tomasoni, Researchers go to School (RgS): how education can meet the scientific innovation, INVO-LEN international conference, 24–25 September 2015, Florence, Italy. In: Ugolini F, Raschi A, Papageorgiou F, editors. Proceedings of the "Innovation in Environmental Education: ICT and intergenerational learning" international conference. Firenze, Italy, 24–25 September 2015, p. 157.

39. DG(SANTE) 2017-6291 Ref. Ares(2017)4789706—02/10/2017 PDF ISBN 978-92-79-52987-0. https://doi.org/10.2875/846869.

40. Von Schomberg R. A vision of responsible innovation. In: Owen R, Heintz M, Bessant J, editors. *Responsible innovation*. London: John Wiley; 2013.

41. Svedin U, editor. New worlds—new solutions. In: *Research and innovation as a basis for developing Europe in a global context, Lund 7–8 July 2009, Sweden*; 2009.

42. Kasperson R, Renn O, Slovic P, Brown HS, Emel J, Goble R, et al. The social amplification of risk: a conceptual framework. *Risk Anal* 1988;**8**:177–87.

43. Judit Papp Komáromi et al. 2010. Training in integrated pest management—No. 2 from Endure. www.endure-network.eu.

44. Michel-Guillou E, Moser G. Commitment of farmers to environmental protection: from social pressure to environmental conscience. *J Environ Psychol* 2006;**26**(3):227–35.

# CHAPTER FIVE

# Practical Implementation of the Principles of the Sustainable Use of Pesticides

## Manpriet Singh*,1, Vasileios P. Vasileiadis*, Anaïs Junger†

*Syngenta Crop Protection AG, Basel, Switzerland
†Syngenta France SAS, Guyancourt Cedex, France
1Corresponding author: e-mail address: manpriet.singh@syngenta.com

## Contents

1. Introduction                                                                      134
2. Policies for Sustainable Agriculture                                              135
   2.1 Sustainable Use of Pesticides in Context of Other EU Policies                 136
   2.2 Key Challenges With Implementing the SUD                                       139
3. EU Policy Design and Enabling Conditions for Implementation                       140
   3.1 Narratives and Networks                                                        141
   3.2 What Motivates Farmers to Change Practice?                                     142
4. Case Studies                                                                       143
   4.1 Case Study 1: Local Diagnosis and Tailored Mitigation Measures to Prevent
       Surface Water Contamination                                                    144
   4.2 Case Study 2: Community Engagement in Managing Remnant and Empty
       Containers                                                                      153
5. Conclusion                                                                         160
References                                                                            162
Further Reading                                                                       164

## Abstract

The Sustainable Use Directive (SUD; 2009/128/EC) is a Community action aiming at the sustainable use of pesticides. It aims at improving use and handling of pesticides, and mitigating human and environmental exposure. Focus areas of the SUD including training pesticide users, inspection of pesticide applicators, and ensuring local processes and infrastructures are in place to manage waste and remnants. The SUD addresses poor use and handling of pesticides on farm to prevent "point source" contamination, e.g., spills or waste water from rinsing pesticide application equipment. In-field measures are also important to mitigate environmental exposure. That is why Member States' National Action Plans also include measures that promote good agricultural practices, tailored to local conditions, to help with preventing "diffuse source" contamination from the field, e.g., runoff. As with the Water Framework Directive (2000/60/EC), the implementation of the SUD is challenged. The implementation across Member States is

*Advances in Chemical Pollution, Environmental Management and Protection*, Volume 2
ISSN 2468-9289
http://dx.doi.org/10.1016/bs.apmp.2018.04.002
© 2018 Elsevier Inc.
All rights reserved.

reported to be inconsistent and the measures implemented not sufficient to deliver improvements in the sustainable use of pesticides. Empirical research about the implementation of the SUD is limited. This chapter presents two case studies that provide insights into how local implementation of sustainable use practices can be improved. The first case study elaborates on the use of digital apps that can both support regional authorities with, for example, setting measurable targets, and farmers with the implementation of tailored mitigation measures on farm and in the field. The second case study proposes the use of actor networks to design context-sensitive approaches to policy implementation.

**Keywords:** Sustainable Use Directive, Policy implementation, Actor network, Digital apps, Point source pollution, Diffuse source pollution

# 1. INTRODUCTION

The Sustainable Use Directive (SUD, 2009/128/EC)[1] is complementary to a number of other EU directives that are meant to achieve a sustainable agricultural sector across all Member States. This is by mitigating any negative impact of food production on human health and the environment. The SUD comprises preventive and precautionary measures to mitigate possible risk and impact related to the use of pesticides. It has seven overarching Community actions for the sustainable use of pesticides: (1) training users, advisors, and distributors on the sustainable and responsible use of pesticides, (2) ensuring the inspection of all pesticide application equipment, (3) prohibiting aerial spraying, (4) limiting the use of pesticides in protected areas (i.e., public areas and conservation areas), (5) disseminating information and raising awareness about potential risks related to the use of pesticides, (6) establishing systems for gathering information on pesticide acute and chronic poisoning incidents, and (7) implementing Integrated Pest Management (IPM) principles and alternative approaches to pesticides. Member States included these Community actions into their National Action Plans (NAPs) which were presented to the European Commission (EC) in 2012,[2] 3 years after the SUD was implemented. In each Member State, different entities have been responsible for drafting the NAPs. This has resulted in each NAP having different priorities, objectives, and methodologies.

In 2017, 5 years after the submission of the NAPs, the DG Health and Food Safety reviewed the progress in implementing the sustainable use of pesticides in Member States. The research was based on a survey conducted among all Member States and fact-finding missions to Denmark, Germany,

Italy, The Netherlands, Poland, and Sweden.[3] The SUD has the potential to deliver on the sustainable use of pesticides; however, the EC concluded that the overall implementation has been "patchy" and progress has been "limited and insufficient to achieve the environmental and health improvements."[4] The EC recommends two key improvements, namely that Member States should make their objectives measurable and that the indicators used to measure progress on reaching the targets should be harmonized across all Member States. The EC also recommends that there should be an increased focus on the implementation of IPM principles, the use of low-risk substances, and the implementation of improved systems to monitor poisoning from pesticides.[5] The overview report, that summarizes the findings from the survey and the fact-finding missions, does not elaborate on the reasons for limited uptake of the SUD's preventive and precautionary measures.

This chapter includes two case studies that address two specific measures for the sustainable use of pesticides, namely (1) measures to protect the aquatic environment and (2) measures for the safe handling and storage of pesticides and treatment of their packaging and remnants. These case studies contribute to a limited amount of empirical research that is available on the practical implementation of sustainable practices including the sustainable use of pesticides. The studies showcase how to implement good agricultural practice effectively and can be used to address specific challenges related to the implementation of the SUD, including: a reported lack of measurable targets, difficulties with compliance monitoring at farm level, farmer incentives to change practice, improving the flow of information to farmers, awareness raising among the farming community and collaboration between key local stakeholders.

Before the two case studies are discussed, the chapter draws a context around the establishment of the SUD and describes how it is linked to other EU policies. The chapter also elaborates on the identified challenges with the implementation of the SUD, and particularly on the importance of participation of local stakeholders, the local context of policy implementation, and farmers' motivation and constraints to change agricultural practices.

## 2. POLICIES FOR SUSTAINABLE AGRICULTURE

Food production is the primary source of income for around 5% of the EU population.[6] The agrofood industry accounts for over 4% of the EU's GDP and employs around one-fifth of its population.[7] Agriculture is thus an important sector for the EU and has to remain competitive, with rural areas that are economically vibrant. However, there are several challenges identified in maintaining agricultural activity and food production. This includes challenges related to a climate with more extremes and a reduced availability of natural resources.[8] Good agricultural practice can help address some of these challenges. The implementation of good agricultural practice and sustainable farming in Europe is mainly driven by social and political concerns around productivity, water quality, and biodiversity. The increased public demand for food produced at farms that have low or no impact on the surrounding environment is as well an important driver.[9–12] Citizens perceive farmers as stewards of the rural environment. They are expected to produce healthy and safe food, while managing the environment and maintaining its aesthetic value.[13] The pull for good agricultural practice goes hand in hand with a policy push for farmers to adopt good agricultural practice. Both the consumer pull and the policy push create opportunities for farmers such as developing new markets for products and services, and providing for public goods (e.g., biodiversity preservation, landscape management).[14]

The practical implementation of EU policies and directives that aim at introducing good agricultural practice has been challenging. The implementation of the SUD shows similar challenges as with the implementation of the Nitrates Directive in 1991 and the Water Framework Directive (WFD) in 2000.[15] Challenges are mostly around (1) the participation of local stakeholders, (2) the adoption of good practice at farm level, and (3) measuring and reporting on these adoption rates.

## 2.1 Sustainable Use of Pesticides in Context of Other EU Policies

The EU lists the following principles as critical for realizing a sustainable agricultural sector: (1) producing safe and healthy food, (2) conserving natural resources, (3) ensuring economic viability of the sector, (4) delivering ecosystem services, (5) managing the countryside, (6) improving quality of life in rural areas, and (7) ensuring animal welfare.[2] Fig. 1 illustrates a timeline with key EU legislation, including the SUD, which has contributed to shaping these principles.

The Birds Directive was the first Community action related to the environment. Initially, there was no link between the Birds Directive and farm management practices. However, this link established after further revisions of the Birds Directive. The protection of birds has become part of the Sixth Community Environment Action Programme focusing on biodiversity in connection to the Habitats Directive. Since rural areas are considered critical

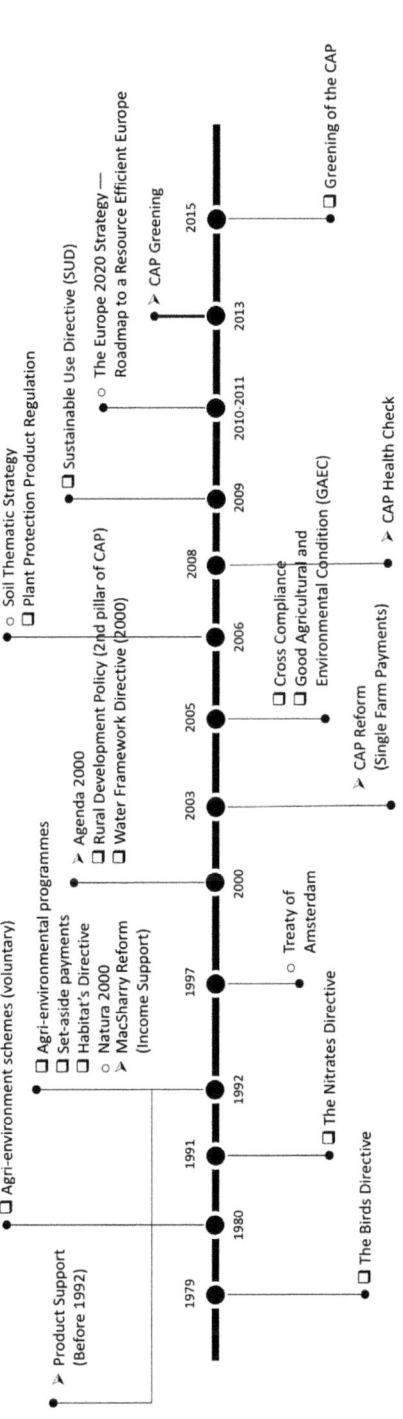

**Fig. 1** EU policies for sustainable agriculture.

in providing and connecting habitats for wildlife which includes birds, there has been a connection of the Birds Directive with farmed areas. The voluntary agri–environmental schemes, first introduced in the 1980s, introduced financial incentives for farmers to manage the rural environment. These early schemes were mainly targeting public concerns related to maintaining the aesthetic value of rural landscapes. They were not focused on addressing challenges to maintain agricultural production into the future, or the prevention of pollution from farms. The Nitrates Directive was the first legislation that specifically aimed at mitigating contamination from agriculture and also promoted good agricultural practice at farm level. It stressed that "agricultural policy must take greater account of environmental policy." The reform and the greening of the Common Agricultural Policy (CAP) did bring agricultural policy closer to environmental policy by making farm support dependent on the farmer respecting and implementing certain practices in support of a healthy rural environment. The CAP is also about a viable agricultural sector as is the regulation on Good Agricultural and Environmental Conditions (GAEC), and the EU's Soil Thematic Strategy. In this context, the SUD is meant to ensure a sustainable and responsible agricultural sector mitigating any possible exposure to or contamination by pesticides. Measures that are part of the SUD can also indirectly address the viability of the agricultural sector as will be demonstrated with the first case study that addresses measures to protect the aquatic environment.

The SUD is complementary to the WFD and vice versa, because both directives aim at reducing environmental exposure. The WFD addresses the ecological, quantitative, and qualitative status of all surface and groundwater bodies. The SUD complements the WFD as its actions can support Member States in reaching a good status of water bodies by, for example, preventing point and diffuse source pollution from farmed areas. The WFD supports the implementation of the SUD as it requires River Basin Authorities (RBAs) draft specific Programmes of Measures (PoMs) for the agricultural sector to prevent water contamination. Some of these measures can help with the implementation of a sustainable use of pesticides on the farm and in the field.

The SUD is different from existing EU directives and regulations that support sustainable agriculture because of its focus on training farmers and preventing operator and bystander exposure. The latter is considered equally important as preventing environmental exposure. A similarity with other directives is that there is no direct financial incentive for users, advisors, and distributors of pesticides to change practice. Farm innovations and farmers' access to information and technologies are critical to realize the objectives of the SUD. The implementation of the sustainable use of pesticides is very much dependent on the availability of local infrastructures and tools, i.e., collection schemes for empty pesticide containers or new pesticide applicators. Implementation is similarly driven by sharing of good practice among local communities depending on identification of current gaps in understanding what assistance local actors require to implement good practice. This can be realized through working across disciplines, linking technical aspects of sustainability to the local social, economic, and political aspects that are key in designing the best and easiest way for farmers to implement good agricultural practice. The importance of sharing information and dialogue between local communities, the availability of enabling infrastructures, and access to the right tools and technologies stress the need of applying context-sensitive approaches to the implementation of the SUD. Addressing dependencies between people, or between people and tools, can help with designing local approaches to address challenges with the on-ground implementation of the sustainable use of pesticides.

## 2.2 Key Challenges With Implementing the SUD

The review performed by DG Health and Food Safety, and a study on the implementation of the provisions of the SUD on IPM report that Member States have taken an embracing, but not completely satisfactory, spectrum of actions to achieve the environmental and health improvements that the SUD was designed to achieve.[3,16] Most Member States have drafted diverse quantitative objectives on risk and/or use reduction and only in five Member States the NAP has high-level measurable targets. Over a period of 5 years, inspection and training and certificate systems were established with, for example, a large number of sprayers being inspected and professional operators being trained and certified. Nevertheless, the level of compliance was found difficult to be assessed due to a lack of reliable data in some Member States. Similarly, the progress on the protection of the aquatic environment or specific areas such as public parks, as well as the safe handling and storage of pesticides and remnants was difficult to be evaluated due to a lack of measurable targets in most NAPs and a lack of farm-level data.[3]

The role of economic instruments in the promotion of the sustainable use of pesticides is also underlined and is strongly encouraged; however, not many Member States have introduced such incentives.[3,16] Financial incentives are available to farmers who invest in buffer zones, low drift nozzles, and on-farm effluent management systems (i.e., remnants from cleaning the sprayer), but not in all Member States.[17] Allocation of regional or local economic resources to enable a close involvement of advisory services for the implementation of good practice is more difficult. The same holds true for promoting applied multidisciplinary research involving diverse stakeholders and sustaining knowledge transfer through training and information sharing.

## 3. EU POLICY DESIGN AND ENABLING CONDITIONS FOR IMPLEMENTATION

EU policy design and implementation has evolved over time and is now more polycentric.[18] This means that there has been a wider involvement of different public stakeholders. Ideally, at Member State level, this means active involvement of all relevant stakeholders.

Stakeholder and public participation in policy design and implementation has been promoted since the establishment of the EU WFD. The WFD supported early public consultations and involvement of all different users in a single catchment for the successful implementation of water management (Preamble, Directive 2000/60/EC). Member States could decide on how the public consultation was shaped. As a result, different approaches have been designed to enable participation and provide for participatory processes. One example is the use of multimedia platforms such as mobile apps that push and pull data and information to and from stakeholders to ensure their participation and feedback in the process of policy making, and implementation.[19] This is especially when stakeholders are geographically separated or when there are too many stakeholders and not everyone can take part in focused meetings. Another example is the use of Fuzzy Cognitive Mapping which is the clustering of different aspects of a current issue identified or perceived by relevant stakeholders and the drawing of links between them to come to consensus on possible solutions and how these solutions could be implemented.[20]

Tools and processes that are meant to enable participation in policy design and implementation shall continuously be evaluated on their effectiveness.[21] Some approaches may not be effective for a stakeholder group or for the identified issue that requires policy intervention. Evaluations can inform the practical implementation of good agricultural practices.

The implementation of the SUD also demands participation between different stakeholders, not only to mitigate environmental exposure but also to mitigate exposure of pesticide users and bystanders. The participatory character of the SUD has not been studied in depth. Insights gained from implementing the WFD can be used to study participatory approaches that are already in place or that are required to improve the implementation of the sustainable use of pesticide. One of the lessons from implementing the WFD is the importance of understanding local networks consisting of people and objects, and the links between them, e.g., the flow of information from one actor to the other or a cause–effect relationship. The decision-making power of people or groups of people within the network is also an important aspect for understanding how local networks function and how they can be exploited for policy implementation.[22]

## 3.1 Narratives and Networks

Before elaborating on networks, it is important that narratives that shape approaches to policy implementation are challenged when EU directives are implemented locally or when new policies are drafted. Molle[23] distinguished three types of concepts that shape policy design and decision making in the water sector: (1) nirvana concepts, (2) narratives or storylines, and (3) models. Nirvana concepts are "concepts that embody an ideal image of what the world should tend to." A narrative is "a story that gives an interpretation of some physical/social phenomena." Models are "based on particular instances of policy reforms or development interventions … and qualify as success stories" (Ref. 23, pp. 132, 136, 138). An example is Integrated Water Resources Management (IWRM) that became a popular approach among policy makers and donor organizations. IWRM promotes the management of water resources at catchment level following participatory principles by involving different stakeholders from different sectors in water planning and decision making. The EU WFD follows the IWRM principles by "planning measures to ensure protection and sustainable use of water in the framework of the river basin" and "[ensuring] the participation of the general public including users of water in the establishment and updating of river basin management plans" (Directive 2000/60/EC). Narratives about participatory approaches to policy implementation suggested cross-collaboration between different stakeholders. However, participation is more complicated and context specific. This was a key lesson from implementing the WFD. Gaps between River Basin Management Plans (RBMPs) and their practical implementation have contributed to weakening the narrative that was repeatedly used to illustrate the successes of stakeholder and farmer engagement in policy design and implementation. Local contexts influence the level of participation between farmers, and between farmers and other stakeholders.

The concepts outlined by Molle are transferrable to policy design and decision making in other sectors. A critical requirement for the practical implementation of the sustainable use of pesticides is participation of local stakeholders and a continuous dialogue between them. For example, farmers require training on the sustainable use of pesticides and an enabling infrastructure to implement good practices such as the collection of empty pesticide containers, the management of remnants, or sprayer inspection. Local participation in setting up these infrastructures is however not guaranteed.

When one narrative becomes weak, a new narrative takes its place, in this case a narrative for other approaches to ensure stakeholder engagement and participation. Drawing actor networks, for example, based on observations can help in understanding which stakeholders participate and engage with each other and which stakeholders do not. This can on its turn help understanding why local stakeholders do not always follow scripts and how their decisions are shaped (Ref. 24, p. 645). The Actor Network Theory (ANT) is not a new concept but it gained more ground in evaluating the implementation of good agricultural practice and the motivation of actors complying with policy frameworks.[25] ANT helps identifying actors that can be people, groups of people, or objects (i.e., nonhumans). It allows to study the connections and relations between these actors and how they influence each other, as well as, processes within the network that they are part of Latour.[25a]

Both narratives and the ANT will be further explored in the case studies.

## 3.2 What Motivates Farmers to Change Practice?

The implementation of good agricultural practice, including the sustainable use of pesticides, is challenging to realize on-farm and in the field. A customer pull and a policy push for sustainable farming practices does not always end in farmers fully adopting these practices.[26] Financial incentives such as subsidies are important, but payments alone also do not result in environmental benefits. For example, agri-environment measures or ecological focus areas selected by farmers for implementation may be isolated from the surrounding rural landscape and hence not deliver the benefits they were intended to.[27,28] Another example is the implementation of sustainable use practices in the farmyard. If there is no supporting infrastructure in place or there is limited information available to farmers about how to use new farm technologies, this may not result in benefits for the environment or for the farm.

The case studies will illustrate that benefits in terms of improved farm management, increased farm productivity, opportunity to differentiate products, and the availability of supporting systems to adopt new practices are as important as financial benefits. A study by Macgregor and Warren[26] supports this observation. They interviewed 30 farms in Nitrate Vulnerable Zones in Scotland and half of these farms reported to have implemented a buffer zone because: (1) it made spray operations easier by complying with local regulations, (2) it enhanced wildlife, (3) there was government funding available, and (4) it prevented erosion. Farmers said that despite the government funding, they spent more to establish and maintain the buffer zones, and they lost productive land. In this case, benefits other than direct financial support were more important to farmers in their decision to establish buffer zones.

The implementation of good agricultural practice requires an improved understanding of the local farming context (i.e., local agronomy, climate, availability of technologies, social context, local policies, etc.) and of farmers' motivation to change practice. Morris[25] argues that studying the farmer's motivation after participating in an agri-environmental scheme is as important as when farmers decide to participate. These insights can greatly support future policy design.

Greiner and Gregg[29] distinguished three motivational factors for farmers to adopt conservation practice: economic or financial motivation, conservation and lifestyle motivation, and social motivation. They also distinguished four factors that make farmers restrain themselves from implementing good agricultural practice: opportunity costs, resource constrains, uncertainty, and lack of industry cooperation. Each farmer is different, e.g., the level of knowledge, past experience, and dependencies on others. That is why each farmer has different incentives to invest time and money in adopting a new farm management practice. Understanding both their incentives and constraints is critical when designing policies and communicating these policies to them. There should be a range of different and tailored value propositions to farmers so that they can see the benefit of implementing a different practice. The private sector has a role to play in supporting policy makers and farmers in adopting sustainable practice, such as enabling farmers' access to new farm technologies, tools, and advisory services that make their farm operations easier and more sustainable.

## 4. CASE STUDIES

DG Health and Food Safety reported six critical gaps in the implementation of the NAPs: (1) their completeness, (2) the inclusion of measurable targets,

(3) the implementation of the eight IPM principles, (4) testing of spray equipment, (5) compliance at farm level, and (6) reporting on progress. Another gap that was identified is the lack of financial incentive for farmers and other stakeholders to adopt good practice or invest in the infrastructure required to ensure a sustainable use of pesticides. Except for the EU 2017 reports that reflect on the implementation of the SUD, practical research including examples of success and failure in implementation is not available, except for the implementation of IPM.

The two case studies described in this chapter are based on farm and in-field observations as well as face-to-face interactions with different stakeholders, but most importantly with farmers. The first case study proposes a new narrative: the use of digital apps to scale up the implementation of good agricultural practices that are critical for the sustainable use of pesticides. The digital application that is introduced in the case study can resolve most of the above-mentioned gaps in the implementation of NAPs and is tackling one particular, but important cause of environmental exposure to pesticides, namely runoff and soil erosion. The second case study is about the implementation of irrigation technologies that are meant to save water. The ANT is used to form an understanding about local dynamics and interactions between actors and applying these to the prevention of point source contamination, which was observed to be a concern in the study area.

## 4.1 Case Study 1: Local Diagnosis and Tailored Mitigation Measures to Prevent Surface Water Contamination

The SUD complements the WFD as it also aims to protect the aquatic environment, but then with a focus on preventing pesticide contamination. There are different pathways by which residues can end up in water bodies, including point sources contamination, spray drift, runoff, leaching, and drainage (Fig. 2). If good practice is not followed, point source contamination from the farmyard, such as spills and remnants, can make up for the largest input of pesticides into surface water bodies.[30] Point sources can easily be prevented by following best management practices (BMPs) when handling and using pesticides. Compared to point source contamination, diffuse source contamination, such as runoff, is more difficult to manage. Runoff occurs when rainfall or irrigation rates are too high to infiltrate into soils or when the soil is too wet. In both situations, water cannot soak into the soil quickly enough.

Mitigating agricultural runoff and soil erosion is an important measure for preventing surface water bodies from pesticide contamination. If not managed, runoff can contribute to around 10% of the total pesticide load

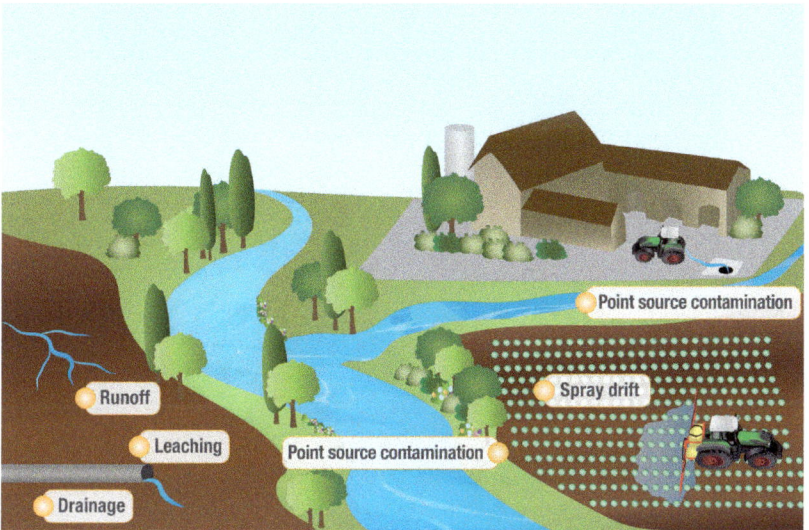

**Fig. 2** Possible sources of water contamination. *Adapted after TOPPS Prowadis (http:// www.topps-life.org).*

entering surface water.[31] Managing runoff is not easy because there can be many different causes to runoff, most of which can only be studied at field level and in discussion with the farmer. A field can be vulnerable to runoff due to steep slopes, adverse slope pattern (i.e., convergent), and adverse soil conditions, i.e., heavy textures, poor structures, and compaction. Management of the crop and the soil is important in determining a field's vulnerability to runoff. Runoff can occur from almost all arable land, even in areas that have a slope of less than 1%.[32]

Twenty NAPs mention the importance of addressing runoff to mitigate environmental exposure to pesticides, especially in protected areas. Buffer strips placed between fields and surface water bodies are referred to as the key measure to mitigate runoff and water contamination. However, in-field measures are reported to be important for vegetative strips to be efficient in buffering runoff and capturing eroded soil. Buffer strips are mainly efficient in capturing eroded soil, which only becomes a main pathway for water contamination when strongly sorbing pesticides are applied on fields vulnerable to runoff.[33] If there is concentrated flow, runoff could pass through the buffer strip (Fig. 3).[34,35]

The effectiveness of in-field measures is also supported by the results of the MARGINS trial site in St. George, which is located to the west of Lake Balaton in Hungary.[36,37] MARGINS stands for Managing Agricultural

**Fig. 3** Runoff passing through a buffer strip in Italy. *Photo credit: Manpriet Singh.*

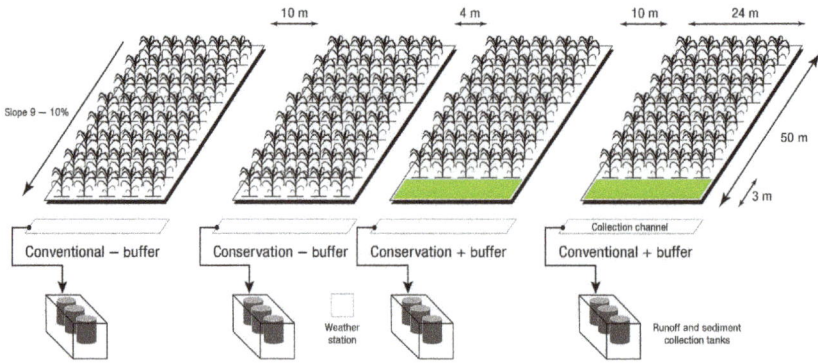

**Fig. 4** MARGINS field trial setup.

Runoff Generation into Surface Waters and has been developed and financed by Syngenta. The site was established in 2009 and consists of four differently managed plots (see Fig. 4). The monitoring is performed by the Hungarian Academy of Sciences. Each plot is connected to runoff and sediment collection tanks. Two plots are plowed and two are managed with crop residues and are minimum tilled (i.e., a tillage technique that does not invert the soil). One plowed and one minimum tilled plot have a 3 m-buffer strip planted at the bottom of the field.

In 2010, 10 runoff events were recorded in St. George. Fig. 5 shows the pesticide runoff from the four different plots. For the minimum tilled plots, pesticide runoff was stopped at source, i.e., in the field. Minimum soil

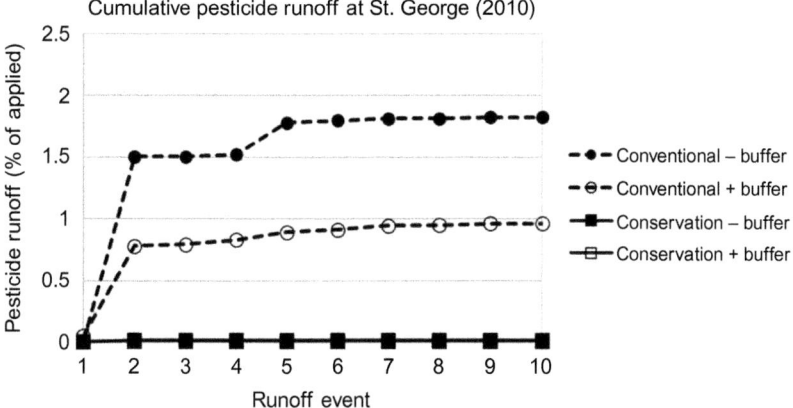

**Fig. 5** Cumulative pesticide runoff at St. George in 2010.

disturbance is thus an important factor for managing agricultural runoff. Buffer strips made a difference, even though in case of the conservation plot, the difference was minimal compared to the small amount of runoff. The buffer strips captured 50%–80% more pesticide runoff when in-field measures (reduced soil disturbance and crop residue management) were implemented.

In-field mitigation measures are included into NAPs and are described in more depth in countries where agricultural runoff is reported to impact farm productivity due to soil erosion and consequent soil degradation, i.e., soil compaction, loss of nutrients, reduced water retention, etc.[38,39] For example, in the United Kingdom the Campaign for the Farmed Environment is part of the SUD implementation and promotes measures such as grass strips across steep slopes or the establishment of a winter cover crops to mitigate runoff. The Hungarian NAP recommends specific tillage practices to maintain healthy soils in areas prone to runoff and soil erosion. In Italy and Spain, the TOPPS Prowadis project is referred to as a voluntary measure to address agricultural runoff. TOPPS stands for Train Operators to Promote best Practices and Sustainability and TOPPS Prowadis is an industry initiative funded by the European Crop Protection Agency. It proposes a general approach to BMPs meant to mitigate agricultural runoff. The BMPs include both buffer strips as well as in-field measures.

In-field measures have another benefit as they can help the farmer manage soil erosion in the field and as such contribute to improving soil health. This is appreciated by farmers who perceive soil erosion to be negative to their farm operations. Controlling soil erosion delivers them direct benefits. For example, in Hungary's rolling landscapes there are many places where

**Fig. 6** Gully formation on a convergent slope, Hungary. *Photo credit: Vince Lang.*

convergent slopes lead to concentrated runoff that result in deep gullies (Fig. 6). This makes it difficult for the farmer to harvest the crop timely. First the farmer will have to harvest the area around the gully after which the gully is disked so that the harvester will not get stuck. Only after this the farmer can start harvesting the whole field. After harvest, and over the winter months, the gully can reappear due to heavy showers.

Managing runoff does not only help the farmer mitigate diffuse source contamination, it also brings benefits to soil health as soil erosion can be prevented and it can become easier to manage fields with farm machinery. However, in practice it is difficult to decide which measures need to be implemented to prevent runoff from vulnerable fields, because each field is different and runoff can be caused by a combination of different factors. Field scouting as well as basic knowledge about soils, crop rotations, and agricultural landscapes is required.

To reduce the complexity of managing runoff and support the scale up of good agricultural practice, Syngenta have developed the Agricultural Runoff and Best Management Practice Tool. This digital tool can be used both by farmers and farm advisors to mitigate pesticide runoff and use plant protection products responsibly and in a sustainable farm environment. The Agricultural Runoff Tool is a hands-on farm support tool that can be used to: (1) identify vulnerable areas to runoff through mapping, (2) diagnose the runoff potential for selected fields using a mobile phone application, and (3) based on this potential, as well as in-field observations supported by pictures and explanations, recommend a set of good agricultural practices (Fig. 7).

**Fig. 7** The three steps in the agricultural runoff and best management practice tool.

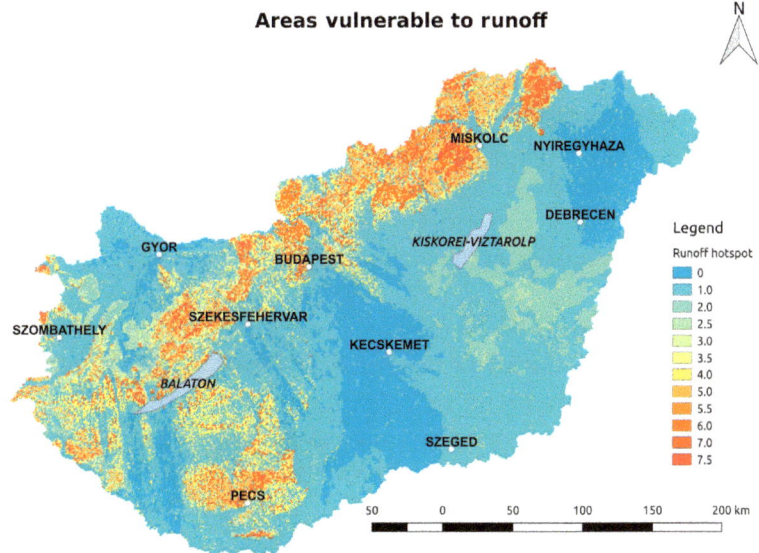

**Fig. 8** Areas vulnerable to runoff in Hungary (map resolution 500 m). *Source: cgiar-csi. org; esdac.jrc.ec.europa.eu; soilgrids.org.*

Maps already available can be uploaded in the online web application and can be consulted offline by the user in the mobile phone application when in the field. Fig. 8 is an example of a runoff potential map for Hungary. The map is an initial diagnosis of runoff potentials based on the average slope, the topsoil permeability, and the depth to a compacted layer.[40] These three factors are inserted into a dashboard with runoff potential scores ranging between 0 for a low runoff potential and 8 for a high runoff potential (i.e., a steep slope, a low topsoil permeability and the presence of a compacted layer close to the soil surface). The runoff potential score from the map can be directly entered into the mobile phone application, or it can be recalculated in the field. By using the mobile application and performing the in-field observations, the runoff potential score is adjusted with ±2. The recommendations to mitigate runoff and soil erosion as well as the in-field observations are summarized into a PDF report that can either be downloaded from the website or send to the user's email account.

The Agricultural Runoff Tool has been piloted in five countries across the EU. The use of the Tool created an understanding of the issue at local scale, which was useful to update the questions in the application for the field diagnosis, and to perform a further refinement of the recommendations for good agricultural practice. Farmers were also asked whether and if so why they are concerned about runoff and soil erosion. These insights supported in further adjusting the tool recommendations. Farmers' feedback on using the Tool (in this case together with a farm advisor) provided insights on whether there is demand for the Tool. For 4 of the 10 pilots, the key cause to runoff, farmer pain points, and farmer feedback is summarized in Table 1.

In summary, the Agricultural Runoff Tool stresses the importance of using pesticides in a sustainable farm environment. The Tool can enable the implementation of practices in support of the sustainable use of pesticides. In-field measures are not only effective in mitigating pesticide runoff and making buffer strips more effective in capturing runoff. In-field measures can also be of interest to the farmer as these can benefit soil health and help maintain farm productivity by preventing soil erosion. The Tool helps scaling up the implementation of good agricultural practice by simplifying an otherwise complex issue. This helps disseminate the know-how around agricultural runoff to nonexperts needed for a wide-scale implementation of good practice. Continuous farmer and farm advisor input can help further tailoring the recommendations included in the Tool as it is adaptable to local conditions.

The use of the application by farm advisors or directly by farmers can contribute to raising their awareness on runoff and soil erosion. Regulatory bodies can push information on good agricultural practice to farmers and request for feedback. Best practices recommended by the Tool can be linked to existing subsidy schemes to provide farmers financial incentives to adopt change practices. Farm reports extracted from the website can also help farmers with negotiating access to certification schemes for good agricultural practice implementation. The latter may be useful especially for high value crops, e.g., vine grapes.

The Tool can deliver useful data. Users can indicate which practices they will consider implementing. The field diagnosis, if repeated after a year, can provide information on the impact of adopting good agricultural practices.

**Table 1** Pilots

| | Hungary, South East of Lake Balaton | Spain, Extremadura | Italy, Vineyard in the North | France, Loire-Atlantique |
|---|---|---|---|---|
| FARMERS | 25 | 8 | 17 | 25 |
| FIELDS | 50 | 17 | 28 | 142 |
| CROPS | Maize, wheat, sun flower, oilseed rape | Maize | Vine grapes | Temporary pasture, maize, cereals |
| CAUSE OF RUNOFF | Most fields have a slope length of >100 m and a convergent slope pattern. Concentrated flow is the key cause for runoff, followed by soil surface compaction | Main cause of runoff is surface infiltration restriction due to soil surface compaction. There are a couple of cases of concentrated flow due to convergent slope patterns | Both surface and subsurface infiltration restriction. When soil in between rows is intensively tilled, subsurface runoff occurs due to compaction at about 30 cm | The main type of runoff is subsurface. However, in some parts of the catchment surface runoff occurs where the slope steepness is $\geq 10\%$ |

*Continued*

Table 1 Pilots—cont'd

| | Hungary, South East of Lake Balaton | Spain, Extremadura | Italy, Vineyard in the North | France, Loire-Atlantique |
|---|---|---|---|---|
| FARMER | – Loss of soil, soil structure, and poor soil water retention<br>– Investments and knowledge is required to change practice<br>– There is an increased time pressure during harvest due to gullies<br>– There is downslope soil accumulation and sedimentation of waterways | Considerable investments are required to change practices and mitigate runoff and soil erosion | – Eroded soil at the bottom of the field needs to be replaced<br>– Eroded farmland does not look good/does not do well to the reputation of the wine producer | – Runoff impacts soil health<br>– Shallow soils and the proximity of fields to water bodies cause soils to be saturated during autumn and winter affecting crops |
| RECOMMENDS TO OTHERS | 7 | 8 | NA | 6–7 |
| FARMER FEEDBACK | – Tool recommendations are based on in-field observations which are critical in understanding the key cause of runoff<br>– The Tool helps in taking management decisions<br>– There is a need to add more detailed recommendations | – Tool helps identifying vulnerable plots and diagnose why they are vulnerable<br>– The one-to-one discussion with the field advisor was appreciated<br>– Farmers learned more about their soil and how this could inform decisions on irrigation scheduling | – The knowledge gained about runoff and soil erosion is valued<br>– Good agricultural practices that support the sustainable production of wine are welcome | – Tool helps identifying most vulnerable plots<br>– The Tool is used by an expert, hence the diagnosis is not done impartially<br>– Recommendations make sense<br>– Readiness to adapt land management, if supported with subsidies |

Recommends to others, i.e., the farmer recommends the Tool to other farmers rated on a scale from 0 to 10 (10 = quite sure, 0 = definitely not).

This is done by recalculating the runoff potential score and also by comparing in-field observations. The use of free satellite data and remote sensing images can potentially also support with the comparison of the situation before and after the implementation of good agricultural practice. Based on farmer data sharing agreements, the Agricultural Runoff Tool can also provide regulatory bodies and other local stakeholders reporting data on the level of compliance with NAPs. Aggregated and anonymized farm and field data can be collected and published under an open data license to demonstrate the use of the Tool on farms as well as for practical research activities. As such, farmers can also be able to see how many other farmers are using the Tool which can trigger farmers' social motivation.

The Agricultural Runoff Tool promotes a new form of participation, engagement, and knowledge and information sharing between stakeholders. The Tool allows for adding a new narrative to the implementation of good agricultural practice, which is the use of self-assessment and field diagnosis tools available on smartphones or tablets to increase on-farm implementation of good agricultural practice. These are easy in use and provide farmers with valuable advice. In-field and digital diagnosis tools support with setting measurable targets for the sustainable use of pesticides and also support data sharing about on-farm implementation of good agricultural practice.

## 4.2 Case Study 2: Community Engagement in Managing Remnant and Empty Containers

This case study outlines the need for farms to adopt sustainable practices and the effectiveness of regional policies and local networks in enabling their implementation in the highly valued environment of the Doñana National Park. The Park, surrounded by a natural pine forest, is located on the right bank of the Guadalquivir River in the province of Huelva in Spain (Fig. 9).

The natural pine forest area is in an important groundwater recharge area of the Doñana aquifer. This is the main source of freshwater flow to the wetlands of the Park. The unconfined part of the aquifer is located in the natural pine forest area which is close to the strawberry production farms. Strawberries are grown on ridges that are draped with black plastic for weed control and moisture management. Irrigation lines are incorporated into these ridges and are also used to apply chemicals (Fig. 10).

Between 1956 and 2007, around 70% of Doñana's former marshland area has been converted into farmland.[41] The upsurge of strawberry farms and the amount of irrigated land are reported to have disturbed the water flow into the aquifer and impacted the wetland's ecosystems. Halting the growth

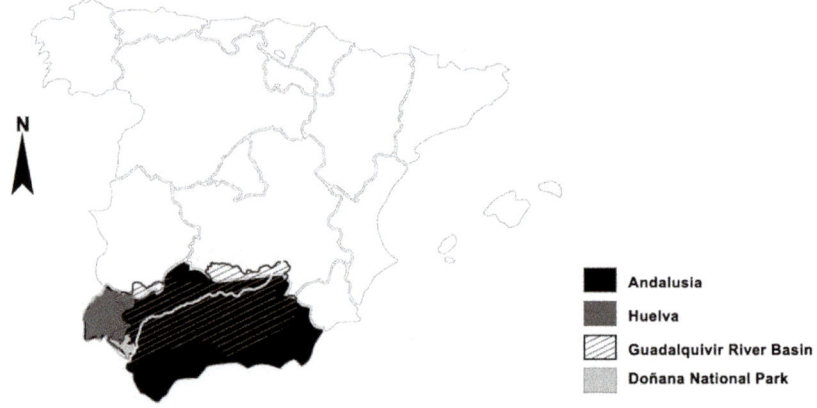

**Fig. 9** The Doñana National Park.

Andalusia
Huelva
Guadalquivir River Basin
Doñana National Park

**Fig. 10** A strawberry farm in Rociana del Condado, Huelva. *Photo credit: Manpriet Singh.*

of farms and restricting the use of water for irrigation has been contested due to the high economic value of strawberry production. In Huelva, around 9500 ha of land is used for berry production (strawberry, raspberry, and blackberry), which is almost 95% of the total production in Spain.[42,43] Huelva exports more than 250,000 tons of berries a year. Due to the continuous increase in market demand, the export value reached about €550 million by 2015.

Farmers have financial incentives to grow and export berries throughout the year.[44,45] Year-round production is possible by introducing an early season, starting in November until April, while the main production season remained from February to July. Year-round production resulted in more use of water and agricultural inputs.

Recent policy interventions have attempted to connect economic development goals in Huelva with conservation goals in view of the reported impact of farming practices on the Doñana wetlands. However, there is no regular dialogue and interaction between the following key stakeholders in the study area: (1) farmers, most of whom are organized in local irrigation communities or a farmer's corporative and who are closest to the management of the strawberry production areas, (2) the Andalusian regional authority who manages the pine forest area and the National Park, and (3) the Guadalquivir RBA who is in charge of implementing the RBPM. The irrigation communities and the farmer cooperatives focus on extracting the highest economic value from agricultural production. Most farmer cooperatives do produce under certification schemes that value sustainable production, e.g., reducing resource input and waste. The Andalusian authority focuses on the implementation of conservation practices to protect the Park's ecosystems. It drafted several regulations that aim at the implementation of sustainable practices in the aquifer's recharge area, prohibiting the further growth of strawberry farms into the pine forest and reducing the use of groundwater for irrigation by investing in modern irrigation schemes.[46,47] The Guadalquivir RBA and the Andalusian regional authority collaborated on reducing the use of groundwater in the recharge area.

The Guadalquivir RBA includes in its RBMP three key measures for the agricultural sector: (1) to convert irrigation systems into drip irrigation systems that have a higher irrigation efficiency and thus can save water, (2) to end overexploitation of groundwater bodies, and (3) to prevent point and diffuse source contamination. In Huelva, irrigation systems have been modernized to prevent overexploitation of groundwater bodies by transferring $4.99\,hm^3$ of surface water from another river basin and renewing the irrigation infrastructure. Next to investments made by the regional authority, EU structural funds have been used to innovate the irrigation sector and farmers also contributed financially. Farmers said to be motivated to cofinance the modernization of the irrigation systems and dismantle their wells when connecting to the new system because of several reasons among which the commitment of their irrigation community and all other member farmers to invest in the project. Farmers also said to invest in irrigation modernization because precision irrigation could help reduce operational costs as well as improve crop health. However, not all farmers in Huelva are member of an irrigation community. One of the reasons is that membership is expensive and small-holders said not to be able to afford it. These farmers are in most cases also not aware of precision irrigation techniques and other farm technologies that can help them improve irrigation scheduling, such as the use of soil moisture probes. This is partly caused by a low level of connection with progressive farmers in the region, and partly by a low level of education. This example shows that membership of irrigation communities, interaction with other farmers in the region, knowledge about precision irrigation technologies, and an under-standing of financial and crop-related benefits of new technologies are the most important reasons for farmers to participate and contribute to the implementa-tion of sustainable water use practices.

These insights on farmer motivation were gained by drawing an actor network based on observations and interviews conducted in the study area (Fig. 11). Morris,[25] for example, applied the ANT to understand farmers' motivation to participate in agri-environmental schemes in England. She argued that ANT offers the possibility to include both nonhuman and human entities into the assessment and draw links between them for a more complete investigation of the social, political, and economic context in which farmers act. Drawing the "water network" for Huelva helped understanding existing relations and missing links between different actors. The water network

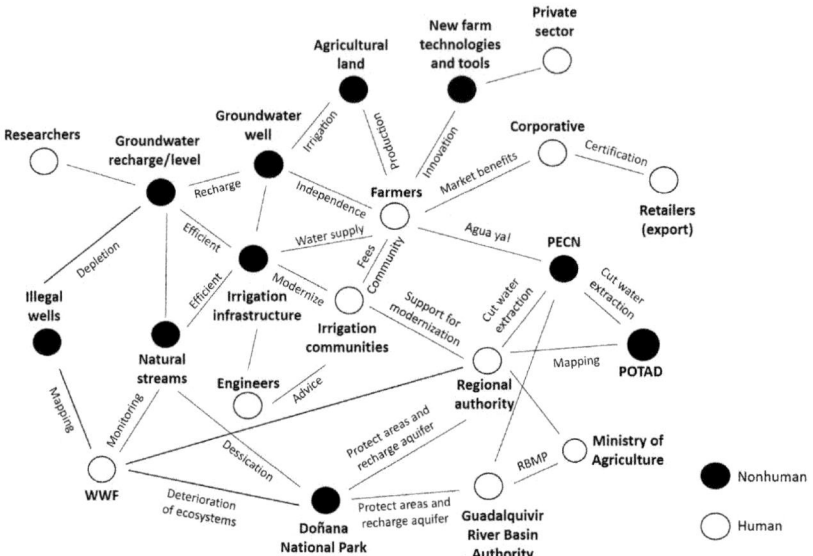

**Fig. 11** Actor network in Huelva.

includes also two key policy actions. PECN is the Special Plan for the Regulation of Irrigation meant to gradually reduce the amount of water allocated to farmers as a result of the irrigation modernization project. POTAD is the Ordination Plan which is a large-scale mapping exercise performed by the regional authority to prevent further expansion of farms into the Natural Park.

Mapping the water network brings insights that are also useful for reviewing the implementation of practices related to the sustainable use of pesticides.

- Most farmers in Huelva are member of local irrigation communities. Irrigation communities or Water User Associations in Spain, as well as in other countries, have social, political, and economic power, e.g., they can influence and are an important stakeholder in local decision making. Member farmers are sure that the irrigation community will protect their rights in front of authorities and other stakeholders. Farmers interviewed in Huelva are said to be "proud" of their membership of the irrigation community and support the irrigation community in its activities. Membership of the irrigation community is also reported as a benefit by farmers because they receive advice on irrigation scheduling from engineers hired by the community.
- There is no link drawn between the regional authority and the Guadalquivir RBA as both do not directly collaborate with each other in the study area. They are connected to each other through the PECN because irrigation modernization, which is promoted by the RBA, is cofinanced by the regional authority. Interviewed farmers and the irrigation community responded that the RBA is not directly active in the area. There is no direct communication or sharing of information with the RBA.
- "Agua Ya!" is a link drawn between farmers and the PECN. It symbolizes farmers' protest against the reduction in the total amount of water allocated to their farms. The PECN regulation is drafted and communicated by the regional authority and is supported by the Guadalquivir RBA. However, the regulation does not directly come from the irrigation communities or is supported by the irrigation communities. This has prompted protests from farmers.
- There is no common understanding and agreement on the status of the Doñana wetlands. The regional authority and farmers claim that there is no issue, while NGOs, who are active in the region, claim that the wetland is continuously deteriorating.

This case study draws a parallel between the sustainable use of water resources and the sustainable use of pesticides. In the strawberry production

**Fig. 12** Disposal of chemical containers, Rio Tinto, Huelva. *Photo credit: Manpriet Singh.*

area, the challenge identified with sustainable use of pesticides is the management of remnants and empty pesticide containers (Fig. 12).

"Remnant" is waste water from cleaning application equipment or from rinsing empty pesticide containers. Remnant management is part of preventing point source contamination from the farmyard. Collecting, rinsing, and safely disposing of empty containers is also part of preventing point source contamination, i.e., preventing leaks and spills from containers. Pesticide containers may not be rinsed properly, because farmers are not aware or they are not lawfully required to rinse containers before disposal. Member States' NAPs underline the importance of managing both remnants and empty pesticide containers. There are Member States that have established container collection schemes and organized for their safe disposal as part of their NAP.[3] Compared to container collection schemes, remnant management systems, i.e., biological, chemical, or physical systems to manage waste water containing pesticide residues, are not widely implemented and are also not required to be available on farm. In Huelva, there are no legal requirement or financial triggers to manage point sources. That is why the implementation of container rinsing and on-farm management of remnants will most likely be triggered by other local incentives.

There are several lessons from the sustainable use of water resources that can be applied to the sustainable use of pesticides in Huelva. First of all, the irrigation communities fulfill an important role in rural areas. They act as social networks as they connect farmers, provide farmers advice related to irrigation practices, and also represent farmers in front of other stakeholders, including authorities. Currently, the community's main focus is on optimizing the irrigation systems. The irrigation communities do not interfere with farm-level management. They also do not provide advice on the sustainable use of pesticides, or any other good agricultural practice not directly related to efficient use of water resources. More active engagement from the irrigation communities could help with the practical implementation of NAPs. Rather than laying the overall responsibility of container management at farm level, irrigation communities may be better positioned to organize or communicate about the need for the collection and safe disposal of empty pesticide containers. The farmers' "Agua Ya!" protest was partly because the irrigation community was not actively communicating about the reduction in the total amount of water allocated to farmers, which is an expected outcome of the irrigation modernization project.

What can also be learned from the actor network drawing is that farmers do not receive any advice, or any support for the implementation of new technologies on their farms that can help them with understanding and implementing precision irrigation, optimizing the use of water at farm level. Large farms have invested in these technologies. Smallholders and also farmers who are not member of an irrigation community have no academic degree in agronomy and do not know how these technologies work and what operational benefits these can bring them. These farmers have not adopted any of these practices at farm level. There is a need for the two key local stakeholders, i.e., the Andalusian regional authority and the Guadalquivir RBA, to better understand what motivates farmers (large and small) to change practices for a sustainable use of pesticides, and initiate a dialogue with farmers to identify any possible gaps that hinder adoption.

Another lesson is the importance of collaboration between the RBA, which is in charge of implementing the EU WFD, and the regional authority who is responsible for the implementation of the NAP. Both regulatory bodies have a common goal, which is protecting the aquatic environment from pollutants. Interaction with farmers through the regional authority can also be further investigated to ensure that the actors in the network are connected to each other and farmers are informed about sustainable use practices. Extending financial support for system modernization to disseminating know-how and enabling a supporting network for container collection could be beneficial.

There is also a role for the private sector. Large commercial farms in Huelva have invested in simple tools such as soil moisture probes or more complicated computer programs and mobile applications to manage irrigation scheduling according to crop need. These commercial farms also participate in certification schemes that promote good agricultural practice. Specialized consultation firms are active in the region to test new irrigation programming systems and support farmers with their implementation. One such system is called Hidrosoph, which involves the farmer installing a weather station, soil moisture probes and flow meters to collect data and insert it into a software program called Irristrat to calculate the crop water requirements and put together the irrigation schedule by themselves based on the most recent data. Further innovations can make low-cost technologies available to smallholders. Low-cost technologies can also support with the prevention of point source pollution from the farm such as remnant management systems. In connection to certification schemes, this can incentivize farmers to follow good practice at farm level.

At last, multistakeholder platforms can help to initiate open dialogue about the status of water bodies and the overall farmed environment. This forum could also serve for any farmer corporative to participate and highlight the needs of farmers.

# 5. CONCLUSION

The SUD was implemented in 2009 and aims at the sustainable use of pesticides to mitigate both human and environmental exposure to pesticides. In 2012, Member States submitted their NAPs for the implementation of the SUD. NAPs were reviewed by the EU in 2017. There were success stories to share, but also areas of improvement.

A lack of measurable targets, farm-level data, tools to track farm-level implementation, and incomplete reporting were among others identified as key areas for improvement. Little participation between local stakeholders, lack of enabling infrastructures, poor access to knowledge, and information and technologies, that can make it easy and attractive for farmers to implement good practice, are the major challenges for the on-ground implementation of the principles of the SUD. Adding to this, farmer motivation, which can be different per individual farmer, is not mapped. Next to compliance to local regulations alone, the intrinsic motivation of local stakeholders is an equally important enabler for farm-level implementation of responsible and sustainable pesticide use practices, and good agricultural practice. All NAPs include measures addressing the key principles of the SUD, yet each NAP has different priorities, objectives, and methodologies. There are NAPs that only focus on the responsible use and handling of pesticides and waste, and NAPs that on top of that include specific measures to ensure that pesticides are used in a sustainable farm environment. The latter can be attractive to farm managers, because good agricultural practice can also benefit soil, crops, and farm management operations such as harvesting.

A limited amount of empirical research is available on the practical implementation of the SUD. Empirical research is a must to test policy narratives and if needed challenge these narratives. Also, empirical research supports with the development of practical tools that can support policy implementation and deliver insights into local dynamics that influence changes in farm and field management. With the use of two case studies that include on-farm and in-field observations and interactions with different local stakeholders, key lessons were extracted that can support the implementation of sustainable use practices. The two case studies address different activities in support of implementing the SUD: in-field measures to prevent diffuse source contamination through runoff and soil erosion mitigation, and on-farm measures to prevent point source contamination through container and remnant management. While the first case study proposed the use of digital tools and mobile apps to ensure that pesticides are used in a sustainable farm environment, the second case study drew a parallel between the efficient use of water resources and the sustainable use of pesticides.

Participation and continuous dialogue between local stakeholders has been identified as a key factor for the implementation of the SUD. A thorough understanding of how different local actors are related to each other is required to design local policy and its implementation, e.g., what connects people, and what separates different regulatory bodies from each other? Also understanding the link between people and objects is critical to identify gaps such as access to pesticide container collection schemes or the availability of remnant management tools. Context-sensitive approaches to policy implementation are recommended.

Another factor is the use of new tools that enable information sharing, training, awareness raising, reporting, and deliver interesting benefits to farmers. Digital apps can help regional authorities with setting measurable targets, while involving people at the microlevel. Tools such as the Agricultural Runoff Tool support data sharing on the implementation of good agricultural practices. The Tool allows local adaptation of recommendations to fit local conditions, addresses local challenges to farming, and triggers farmer incentives (not only financial incentives) to implement new ways of working.

Finally, there is a role for the private sector, namely to enable uptake of good practice, supporting policy makers and farmers. Developing and piloting farm management solutions and innovations together with farmers enables to deliver tools to the market that can serve multiple purposes, e.g., from tools that distributors or others can use to deliver farmers with advisory services, to tools that can make farm operations easier, more sustainable, and can as well provide farmers with opportunities to develop new markets for their products.

## REFERENCES

1. Sustainable Use Directive (SUD) 2009/128/EC of the European Parliament and the Council of 21 October 2009 establishing a framework for Community action to achieve the sustainable use of pesticides.
2. European Union. Sustainable agriculture for the future we want. European Commission 2012.
3. European Union. Overview report sustainable use of pesticides. DG Health and Food Safety 2017.
4. COM Press Release. Progress report on NAP of EU MS under the sustainable use directive (SUD). October 2017.
5. DG Health and Food Safety. Update on developments on the sustainable use of pesticides (Directive 2009/128/EC). European Commission 2017.
6. European Union. How many people work in agriculture in the European Union: an answer based on Eurostat data sources. European Agricultural Economics Briefs 2013.
7. Moussis N. *Access to European Union: law, economic, policies. The ultimate textbook on the European Union.* UK: Intersentia Ltd; 2011.
8. FAO. The future of food and agriculture—trends and challenges. Rome 2017.
9. European Commission. Attitudes of European citizens towards the environment. Special Eurobarometer 295/Wave 68.2. TNS Opinion & Social 2008.
10. European Commission. Roadmap to a resource efficient Europe. Communication from the Commission to the European Parliament, the Council, the European Economic and Social Committee and the Committee of the Regions Brussels COM 2011: 571, https://doi.org/10.3389/fimmu.2018.00571.
11. Vermeir I, Verbeke W. Sustainable food consumption: exploring the consumer 'attitude–behavioral intention' gap. *J Agric Environ Ethics* 2005;**19**:169–94.
12. Weatherell C, Tregear A, Allinson J. In search of the concerned consumer: UK public perceptions of food, farming and buying local. *J Rural Stud* 2003;**19**(2):233–44.
13. Hall C, McVittie A, Moran D. What does the public want from agriculture and the countryside? A review of evidence and methods. *J Rural Stud* 2004;**20**:211–25.
14. Singh M, Marchis A, Capri E. Greening, new frontiers for research and employment in the agro-food sector. *Sci Total Environ* 2014;**472**:437–43.
15. Water Framework Directive (WFD) 2000/60/EC of the European Parliament and the Council of 23 October 2000 establishing a framework for Community action in the field of water policy.
16. Movses G, Micheli Marie-Antoinette. Study on the implementation of the provisions of the sustainable use directive on integrated pest management: experience of eleven member states of the European Union. 13 November 2015.
17. COM Press Release. On member state national action plans and on progress in the implementation of Directive 2009/128/EC on the sustainable use of pesticides. October 2017.
18. Pahl-Wostl C. Participative and stakeholder-based policy design, evaluation and modeling processes. *Integr Assess* 2002;**3**(1):3–14.

19. Pereira AG, Rinaudo J, Jeffrey P, Blasques J, Quintana SC, Courtois N, et al. ICT tools to support public participation in water resources governance & planning: experiences from the design and testing of a multi-media platform. *JEAPM* 2003;**5**(3):395–420.
20. Mouratiadou I, Moran D. Mapping public participation in the Water Framework Directive: a case study of the Pinios River Basin, Greece. *Ecol Econ* 2007;**62**(1):66–76.
21. De Stefano L. Facing the water framework directive challenge: a baseline of stakeholder participation in the European Union. *J Environ Manage* 2010;**91**:1332–40.
22. Benson D, Fritsch O, Cook H, Schmid M. Evaluating participation in WFD river basin management in England and Wales: process, communities, outputs and outcomes. *Land Use Policy* 2014;**38**:213–22.
23. Molle F. Nirvana concepts, narratives and policy models: insights from the water sector. *Water Altern* 2008;**1**(1):131–56.
24. Mosse D. Is good policy unimplementable? Reflections on the ethnography of aid policy and practice. *Dev Change* 2004;**35**(4):639–71.
25. Morris C. Networks of agri-environmental policy implementation: a case study of England's Countryside Stewardship Scheme. *Land Use Policy* 2004;**21**:177–91.
25a. Latour B. On actor-network theory: a few clarifications. *Soz Welt* 1996;**47**:369–81.
26. Macgregor CJ, Warren CR. Adopting sustainable farm management practices within a Nitrate Vulnerable Zone in Scotland: the view from the farm. *Agric Ecosyst Environ* 2006;**113**:108–19.
27. European Commission. Review of greening after one year. 2016.
28. IEEP. *Green direct payment: implementation choices of nine Member States and their environmental implications.* Institute for European Environmental Policy (IEEP); 2015.
29. Greiner R, Gregg D. Farmers' intrinsic motivations, barriers to the adoption of conservation practices and effectiveness of policy instruments: empirical evidence from northern Australia. *Land Use Policy* 2011;**28**:257–65.
30. Müller K, Bach M, Hartmann H, Spiteller M, Frede HG. Point and non-point source pesticide contamination in the Zwester Ohm Catchment (Germany). *J Environ Qual* 2002;**31**(1):309–18.
31. Neumann M, Schulz R, Schäfer K, Müller W, Mannheller W, Liess M. The significance of entry routes as point and non-point sources of pesticides in small streams. *Water Res* 2002;**36**:835–42.
32. Vymazal J, Březinová T. The use of constructed wetlands for removal of pesticides from agricultural runoff and drainage: a review. *Environ Int* 2015;**75**:11–20.
33. Leonard RA. Movement of pesticides into surface waters. In: Cheng HH, editor. *Pesticides in the soil environment: processes, impact, and modeling. SSSA Book Series*, vol. 2. Soil Science Society of America, Inc.; 1990. p. 303–50.
34. Arora K, Mickelson SK, Baker JL. Effectiveness of vegetated buffer strips in reducing pesticide transport in simulated runoff. *Trans ASAE* 2003;**46**:635–44.
35. Verstraeten G, Poesen J, Gillijns K, Govers G. The use of riparian vegetated filter strips to reduce river sediment loads: an overestimated control measure? *Hydrol Process* 2006;**20**:4259–67.
36. Jakob G, Madarász B, Szabó JA, Tóth A, Zacháry D, Szalai Z, et al. Infiltration and soil loss changes during the growing season under ploughing and conservation tillage. *Sustainability* 2017;**9**(1726):1–13.
37. Madarász B, Bádonyi K, Csepinszky B, Mika J, Kertész A. Conservation tillage for rational water management and soil conservation. *Hung Geogr Bull* 2011;**2**:117–33.
38. Bakker MM, Govers G, Jones RA, Rounsevell MDA. The effects of soil erosion on Europe's crop yields. *Ecosystems* 2007;**19**:1209–19.
39. Stoate C, Báldi A, Beja P, Boatman ND, Herzon I, Van Doorn A, et al. Ecological impacts of early 21st century agricultural changes in Europe—a review. *J Environ Manage* 2009;**91**:22–46.

40. Panagos P, Borrelli P, Poesen J, Ballabio C, Lugato E, Meusburger K, et al. The new assessment of soil loss by water erosion in Europe. *Environ Sci Policy* 2015;**54**:438–47.
41. Palomo I, Martin-López B, Zorrilla-Miras P, Del Amo GD, Montes C. Deliberative mapping of ecosystem services within and around Doñana National Park (SW Spain) in relation to land use change. *Reg Environ Change* 2013;**14**(1):237–51.
42. Aldaya MM, Novo FG, Ramón Llamas M. Incorporating the water footprint and environmental water requirements into policy: reflections from the Doñana Region (Spain). In: Martínez-Cortina L, Garrido A, López-Gunn E, editors. *Re-thinking water and food security*. CRC Press; 2010. p. 193–217.
43. USDA FAS Gain Report, Global Agricultural Information. Spanish soft fruit production increases and consolidates as an alternative to strawberry. Gain Report Number SP1604; 2 October 2016.
44. FEPEX. *Federación Española de Asociaciones de Productores Exportadores*. http://www.fepex.es/datos-del-sector/ [Accessed 12 June 2016].
45. ICEX. *España Exportación e Inversiones*. http://www.icex.es/icex/es/navegacion-principal/todos-nuestros-servicios/informacion-de-mercados/estadisticas/sus-estadisticas-a-medida/estadisticas-espanolas-estacom/index.html; 2016 [Accessed 12 June 2016].
46. Martín-Lopéz B, García-Llorente PI, Montes C. The conservation against development paradigm in protected areas: valuation of ecosystem services in the Doñana social-ecological system (southwestern Spain). *Ecol Econ* 2011;**70**(8):1481–91.
47. Zorrilla-Miras P, Palomo I, Gómez-Baggethun E, Martín-López B, Lomas LP, Montes C. Effects of land-use on wetland ecosystem services: a case study in the Doñana marshes (SW Spain). *Landsc Urban Plan* 2014;**122**:160–74.

## FURTHER READING

48. European Union. Directive 2009/128/EC of the European parliament and of the council.

CPI Antony Rowe
Chippenham, UK
2018-06-07 22:40